Transport Processes in Nature

Propagation of Ecological Influences through Environmental Space

This book provides a new and general perspective on how events or conditions in environmental space have influences elsewhere. In the first part of the book, the authors introduce the general question of propagation of ecological influences through environmental space (terrestrial, aquatic, and aerial) and then present a system for its analysis by organization into four components: initiating events or conditions, vectors conducting influences over space, entities that are transported, and the consequences of these propagation processes. Each of these components is rich in variation and logical complexity. Methods for representing environmental heterogeneity and for modeling transport processes are discussed in the context of such propagations. In the second part, properties of eight general transport vectors and examples of transport models in realistic ecological situations are explained. For each of the vectors, a simulation model linked with ArcView$^{®}$ geographical information system (GIS) is provided on a CD included with the book (users require access to ArcView$^{®}$ GIS software). Although set in an ecological context, the concepts presented will also be of direct relevance to a range of disciplines within the earth and atmospheric sciences.

BILL REINERS is a professor in the Departments of Botany and Geography at the University of Wyoming, and Director of the Wyoming Geographic Information Science Center. His scientific career spans plant ecology, ecosystem ecology, and biogeochemistry. With the development of new technologies for spatial studies, his work now focuses on the consideration of ecological processes in explicit spatio-temporal contexts.

KEN DRIESE is a research scientist at the Wyoming Geographic Information Science Center at the University of Wyoming. He is currently building a remote sensing program at the Center and conducting research on the use of remotely sensed data to help to solve spatially explicit ecological problems. His scientific career has focused on remote sensing and GIS as tools for ecological analysis.

Cambridge Studies in Landscape Ecology

Series Editors
Professor John Wiens *Colorado State University*
Dr Peter Dennis *Macaulay Land Use Research Institute*
Dr Lenore Fahrig *Carleton University*
Dr Marie-Josee Fortin *University of Toronto*
Dr Richard Hobbs *Murdoch University, Western Australia*
Dr Bruce Milne *University of New Mexico*
Dr Joan Nassauer *University of Michigan*
Professor Paul Opdam *ALTERRA, Wageningen*

Cambridge Studies in Landscape Ecology presents synthetic and comprehensive examinations of topics that reflect the breadth of the discipline of landscape ecology. Landscape ecology deals with the development and changes in the spatial structure of landscapes and their ecological consequences. Because humans are so tightly tied to landscapes, the science explicitly includes human actions as both causes and consequences of landscape patterns. The focus is on spatial relationships at a variety of scales, in both natural and highly modified landscapes, on the factors that create landscape patterns, and on the influences of landscape structure on the functioning of ecological systems and their management. Some books in the series develop theoretical or methodological approaches to studying landscapes, while others deal more directly with the effects of landscape spatial patterns on population dynamics, community structure, or ecosystem processes. Still others examine the interplay between landscapes and human societies and cultures.

The series is aimed at advanced undergraduates, graduate students, researchers and teachers, resource and land-use managers, and practitioners in other sciences that deal with landscapes.

The series is published in collaboration with the International Association for Landscape Ecology (IALE), which has Chapters in over 50 countries. IALE aims to develop landscape ecology as the scientific basis for the analysis, planning and management of landscapes throughout the world. The organization advances international cooperation and interdisciplinary synthesis through scientific, scholarly, educational and communication activities.

Also in the series:
J. Liu and W. W. Taylor (eds.) *Integrating Landscape Ecology into Natural Resource Management*
R. Jongman and G. Pungetti (eds.) *Ecological Networks and Greenways*

WILLIAM A. REINERS
UNIVERSITY OF WYOMING, LARAMIE, WYOMING, USA

KENNETH L. DRIESE
UNIVERSITY OF WYOMING, LARAMIE, WYOMING, USA

Transport Processes in Nature

Propagation of Ecological Influences through Environmental Space

PUBLISHED BY THE PRESS SYNDICATE OF THE UNIVERSITY OF CAMBRIDGE
The Pitt Building, Trumpington Street, Cambridge, United Kingdom

CAMBRIDGE UNIVERSITY PRESS
The Edinburgh Building, Cambridge CB2 2RU, UK
40 West 20th Street, New York, NY 10011–4211, USA
477 Williamstown Road, Port Melbourne, VIC 3207, Australia
Ruiz de Alarcón 13, 28014 Madrid, Spain
Dock House, The Waterfront, Cape Town 8001, South Africa

http://www.cambridge.org

First published 2004

Printed in the United Kingdom at the University Press, Cambridge

Typeface Lexicon 10/14 pt.   *System* LATEX 2_ε [TB]

A catalog record for this book is available from the British Library

Library of Congress Cataloging in Publication data

Reiners, William A.
Flows and movements in nature: propagation of ecological influences through
environmental space / William A. Reiners and Kenneth L. Driese.
    p. cm. – (Cambridge studies in landscape ecology)
Includes bibliographical references (p.).
ISBN 0 521 80049 8 (HB) – ISBN 0 521 80484 1 (paperback)
1. Ecology. I. Driese, Kenneth L. II. Title. III. Series.
OH541.R393 2004
577 – dc22 2003064192

ISBN 0 521 80049 8 hardback
ISBN 0 521 80484 1 paperback

Contents

Preface

This book evolved from the conjunction of different experiences, world views, and ontologies of people and disciplines. These conjunctions started with the authors. Reiners received a classical plant ecology education in the early 1960s, which was modified over the decades by his involvement with ecosystem ecology, biogeochemistry, landscape ecology, remote sensing, GIS, and earth system science. Driese was formally educated in the early 1980s and 1990s in the broader environmental sciences, which provided him with a stronger background in mathematics and physical science. The authors met in 1987 and have been collaborators since then.

Other conjunctions influenced the way we came to perceive nature. Reiners taught and practiced conventional ecology from the 1960s through the mid-1990s by focusing on single points in space but with changing properties in time. However, this geographically fixed ontology did not satisfactorily accommodate the author's observations of the world around. It provided no "place" for how polluted air transported from the midwestern USA to the mountain ecosystems of New England deposited acidity via cloud droplets, rain, and snow. Nor did it explain how this acidic precipitation then flowed in complicated, spatially distributed ways through canopies, soils, rocks, watersheds, and rivers. It did not satisfactorily account for the fact that these same forests had to be interpreted in terms of the historic New England hurricane of 1938, a disturbance that originated somewhere west of Africa, rolled up the Atlantic coast, and smashed its way into the heart of New England. Similarly, this spatially fixed viewpoint was disconnected from our understanding of how capital and markets in Europe and North America could be expressed as the conversion of rain forests to banana plantations. In Wyoming, we witnessed the extensive forest fires of 1988 in Yellowstone National Park, debris flows in the Teton Range, dispersal of rabid skunks through river valleys, and redistribution of snow by wind in shrub steppes. These phenomena made it obvious to us that many of the more important ecological phenomena involved propagation of cause from one locus to give rise to an effect elsewhere; transport and location were not suitably incorporated into conventional ecological thinking. There was a disconnection between point-oriented ecology and important propagations of many kinds. Naturally, we knew that inputs and outputs were recognized as properties of point models, but we came to see that the transport processes themselves were mainly subjects of other disciplines such as hydrology and epidemiology while being just peripheral to mainstream ecological thought.

As we explored the nature of propagations, we found that the sciences explaining transport mechanisms were often mature but dispersed among disciplines such as physical oceanography, geomorphology, and micrometeorology. Consolidating that dispersed information into an ecological viewpoint required many more conjunctions.

This book, then, results from bringing together different ways of seeing and explaining nature. In its simplest form, our goal has been to fuse these conjunctions to achieve a confluence between an ontology based on typal systems treated as homogeneous units (points) and an ontology based on nature as fields of movements and flows through heterogeneous environments. We believe such an ontologic confluence will enrich and empower environmental science in general and ecology in particular.

The contents are divided into two parts. The first provides the philosophical and conceptual grounding for propagation of ecological influences. It also describes how propagations can be abstracted and modeled with spatially distributed modeling tools. Because it is efficient to describe propagations in terms of vectors, fuller descriptions of transport processes are organized by vector in Part II. Each chapter defines and describes the physics and biology of a particular vector, shows how appropriate disciplines have developed methods for modeling the vector, and finally provides a simple model for readers to use. The models illustrate how altering primary variables can vary outcomes of transport processes. The purpose and organization of Part II are more fully explained in Ch. 5.

This book will serve as a primary resource for thoughtful ecologists willing to examine their discipline in a fresh and challenging way. We intend it to facilitate the development of broader and more flexible ways of teaching and practicing ecology. We hope that it will bring geography into more intimate connection with ecology and beyond its present deployment for biogeographic issues. Inasmuch as there are no courses on ecological propagations, this book is not designed as a text; rather we hope it will be a source to be read by individuals or discussed in intimate seminar settings.

Acknowledgements

The conceptual development and early phases of production of this book were supported by a grant from the Andrew W. Mellon Foundation. The authors thank William Robertson IV, Program Director, for his belief and support in this project. We are deeply grateful to the National Center for Ecological Analysis and Synthesis (NCEAS) for provided funding for Reiners to develop and write Part I and to outline Part II.

Several members of the Department of Geography at the University of California-Santa Barbara were especially influential in the development of this book. We thank Michael F. Goodchild for his overall support for the vision of this project, and for his special guidance on GIS and environmental modeling issues. Helen Couclelis graciously advised us on Ch. 3, particularly on the philosophy of space and time and on cognition of the environment. Matthew J. Ungerer was especially helpful with GIS data modeling and process modeling, as discussed in Ch. 4. Keith C. Clarke introduced us to the fire model featured in Ch. 9.

Colleagues at the University of Wyoming contributed in various ways. A group of thoughtfully creative graduate students provided a critical forum for fleshing out the dimensions of ecological propagations in the early stages of our thinking. Dennis H. Knight provided insight, stimulation, and encouragement on this topic, which shaped the book in major as well as subtle ways. Stephen T. Jackson provided a critical review of our treatment of wind, and Christopher A. Hiemstra generously gave us his adaptation for snow transport by wind and other background material on wind transport. Philip Polzer programmed the initial animal movement model and designed many of the model interfaces. Gary Beauvais of the Wyoming Natural Diversity Database (WYNDD) provided data for the animal movement model and Douglas A. Keinath, also with WYNDD, provided critical reviews of Ch. 11. David Rush of the Wyoming Geographic Information Science Center wrote the program modifications needed for the fire model presented in Ch. 9. Norma M. Reiners and Ellen V. Axtmann helped with vital tasks of figure generation, literature acquisition, and editing.

This book would not have been possible without the cooperation of a number of scientists who generously allowed us to adapt models they had published in other formats. Bill Massman of the USDA Forest Service gave us permission to use the model illustrating diffusion. Bill also gave us an expert review of Ch. 6. The colluvial–fluvial model adapted in Ch. 7 was provided by Lee Benda and Daniel Miller of the Earth Systems Institute, Seattle, WA. Glen Liston of Colorado State University and Matt Sturm of the US Army are the sources of the original model for wind transport of snow described in Ch. 8. Keith Clarke, University of California-Santa Barbara, generously shared the code for the fire model in Ch. 9. Baxter E. Vieux and Fekado G. Moreda of Texas A&M University provided the demonstration form for their fluvial model described in Ch. 10. Kirsten Parris, of the University of Melbourne, did much of the original research on the propagation of sound and light and is the author of the spring peeper model described in Ch. 13. Dr Parris also added extensively to the chapters on sound and electromagnetic radiation. These contributors also provided critical reviews of individual chapters. We also wish to cite Carol Stevens, Wade Griffith, and Ryan Taylor for their individual contributions to graphic design of original figures.

We thank Alan Crowden of Cambridge University Press for his patient support and faith in this book.

PART I

Flows and movements in ecology

Soil health and water health are not two problems, but one. There is a circulatory system of food substances common to both, as well as a circulatory system within each. The downhill flow is carried by gravity, the uphill flow by animals.

Leopold 1941

1.1   Nature's telegraphy: three anecdotes

A wood thrush lands on an arrow-wood bough, listens, and then sends out his fluty challenge. Atmospheric vibrations pulse outward, bearing his message through the leafy forest canopy. Across the cove, another male thrush hears the message and flies away.

A leaf falls from a chestnut oak on the steep slopes of a West Virginia mountain and joins thousands of others in the deep bed below. Wind blows through the defoliated November forest and the russet leaves drift downslope (Fig. 1.1). Snow comes and, in April, meltwater washes leaves further downslope, some moving into the fast-moving stream cutting across the base of the slope. These become organized into a leaf pack and then reorganized as fungal mycelia and, finally, as black fly biomass. With each transformation, the substance of these leaves, assembled the previous spring in the canopies upslope, moves ever further downstream (Orndorf and Lang, 1981; Kooser and Boerner, 1985).

The floods recede and dust is blown aloft from China's Yellow River basin. The dust is carried eastward across the Pacific where some is deposited over nutrient-depleted waters and onto a volcanic island chain. The dust dissolves in seawater, supplying iron to the otherwise iron-limited plankton, and energy flow through the food chain accelerates (Fung *et al.*, 2000). Dust deposited on old lava flows of the volcanic islands also contributes to the phosphorus supply of aged soils, allowing them to maintain high levels of primary productivity (Chadwick *et al.*, 1999).

These three anecdotes illustrate how phenomena occurring at one location influence phenomena elsewhere in environmental space. What are these influences? How are they propagated across landscapes? How can we understand and predict these propagations? These are the questions addressed by this book.

FIGURE 1.1
Transport of leaf litter from trees to the ground by gravity, thence downslope by running
snowmelt water, and finally into a mountain stream where litter debris is organized into
leaf packs and then further reorganized into finer particles and animal and fungal biomass.

1.2 Propagation of influences across space: views from different subdisciplines within ecology

As the table of contents of every ecology textbook reveals, ecology can be examined at different levels of organization – individual or species, population, community, ecosystem and global – or from different points of view, such as landscape ecology. For the most part, contemporary ecologists specializing within each of these subdisciplines realize the importance of flows and movements when addressing conceptual questions or dealing with specific situations. Typically, however, these realizations are limited to a few kinds of flow and often include a relatively limited understanding of processes underlying these flows. Even more often, ecologists are uncertain as to how such flows can be formally incorporated into their work. These conditions set the stage for a more focused analysis of flows and movements. The following sections describe how the generally recognized subdisciplines composing ecology have come to incorporate the importance of flows and movements in their conceptualization of how nature works.

1.2.1 Ecology at the individual organism level

The study of individual organisms can be described as "autecology," "physiological ecology," or "ecophysiology" of plants or animals, or, depending on taxonomic focus, as "animal behavior." Ecophysiology involves the interactions between plants and animals with fluxes of radiant energy from the sun, sky, and solid objects as well as the flows of wind and water and exchanges with the atmosphere, soil, or water medium (Campbell, 1977; Monteith and Unsworth, 1995). Gas exchange by plants is based on movements by diffusion and mass flow driven by convection processes (Larcher, 1995). Heat regulation by animals is driven by interactions between animal behavior, radiant energy flux, and wind. For animal behaviorists, communication by sound, sight, and even electromagnetic fields – all influences propagated across space – is central to the interpretation and understanding of behavior (Bradbury and Vehrencamp, 1998). To a large degree, organismal ecology is about how organisms cope with the fluxes of abiotic and biotic materials around them.

1.2.2 Population ecology

Early ecologists held "spatially aware" views of nature, but in the middle of the twentieth century, classical population models came to be presented by academics as abstract, isolated systems (Wiens, 2000). Immigration, emigration, and food imports into or out of the defined systems were conveniently defined as either constant and equal or negligible and irrelevant (Pickett and Cadenasso, 1995). Population models were essentially point models in a spatial sense. Range managers and wildlife and fisheries biologists learned from these models but had to manage real populations in actual landscapes and seascapes. In so doing, they took spatial movements into account. Huffaker's microcosm experiments (1958) showed how heterogeneity in environments was important to the regulation and temporal dynamics of populations. Since then, the importance of movements within heterogeneous habitats has become the dominant view of population biology, as illustrated by Paine (1966), Caswell (1978), Paine and Levin (1981), Roughgarden *et al.* (1987), Holt (1993), McLaughlin and Roughgarden (1993), Kareiva and Wennergren (1995) and Hansson *et al.* 1995. Today, the conveniently isolated population is no longer the elegant ideal in population ecology (Rhodes and Odum, 1996; Wiens, 2000). Population ecologists are well aware of the limitations of their models, and spatial interactions have become an intrinsic part of the modeling process. The current situation is well summarized by Polis *et al.* (1997):

> Ecologists are now aware that dynamics are rarely confined within a focal area and that factors outside a system may substantially affect (and even dominate) local patterns and dynamics. Local populations are linked closely with other populations through such spatially mediated interactions as source-sink and metapopulation dynamics, supply-side ecology, and source pool-dispersal effects.

The movements of resources and other species to or from local populations, or the movement of individuals between local populations, are examples of propagated ecological influences through space.

Another area of population ecology – population genetics/microevolution – combines the demographic aspects of populations with gene flow within and between them. Populations possess collective genomes subject to dynamics linked with demographic and dispersal processes. The differentiation of, and gene flow between, local populations is central to the understanding and prediction of species fitness and evolution. Spatial heterogeneity and factors limiting gene flow underlie these processes and are intrinsically geographical in nature. Parts of this research field have recently become embodied in metapopulation biology, an approach that addresses both demographic and genetic issues. Metapopulation biology attempts to quantify gene flow between constituent local populations to assess their openness to, or isolation from, the species at large (Hanski, 1998, 1999). Gene flow is another example of the propagation of ecological influences through space via transmission of spores, pollen, fruits, seeds, and dispersing animals.

1.2.3 Community ecology

Many ecologists focus at the community level: assemblages of species populations co-occurring in space and time. The community concept had a botanical origin but modern activity is largely centered on animal population interactions (Brown, 1995). As such, many of the viewpoints outlined above under population ecology have transferred to community ecology. These viewpoints include a modern appreciation of movements and flows across space (Roughgarden *et al.*, 1987; Holt, 1993; McLaughlin and Roughgarden, 1993; Fahrig and Merriam, 1994; Tilman *et al.*, 1994; Kareiva and Wennergren, 1995). The importance of organism movement into and out of communities is so well recognized that Brown and Gibson (1983) suggested that ecologists imagine, as a null hypothesis, the consequences of a local community being enclosed within a dispersal-proof fence or invisible "force field." The implausibility of such a situation makes apparent the pervasive importance of movements in nature (Holt, 1993).

Movements of abiotic substances are also recognized as important at the community level. Roughgarden *et al.* (1987) make the case with the following statement:

> In both population and community ecology it is increasingly clear that data taken within a study site have limited power to explain what happens in the site. At some scale most ecological systems are open systems, and the control exerted by physical transport processes on population and community dynamics matches the effect of local processes, such as predation and competition among the residents of the site.

Later in this article, they say,

> Ecological studies, except those on very long-lived organisms, tend to regard the earth sciences at arm's length; earth-science processes merely set the backdrop for ecological processes, but do not play a leading role. But earth-science processes actually control both the population and community ecology of the intertidal zone and the community ecology of Caribbean islands to an extent that matches the effect of any biological mechanisms that operate within these ecological systems.

1.2.4 Ecosystem ecology

The original definition of ecosystem was an interacting "system" of the biota (the community) and the abiotic variables with which the biota interact. Tansley coined this term in 1935 in order to resolve ecological community concepts of the time. In historical fact, the "system" concept blending together physical and biological variables was so useful that it led to the development of a new branch of ecology. This branch developed major differences in its culture and style compared with population and community ecology, as well as coming to possess, within itself, a broad variety of subdivisions in terms of foci and approaches (McIntosh, 1985; Chapin *et al.* 2002).

This range of meanings and implications in the term "ecosystem" was complicated further by the adoption of the term into the lexicon of general society. As this happened, the term became corrupted by implications designed to rally or inflame readers and listeners. The word "ecosystem" cannot be spoken without stimulating antisocial expressions of anger in certain public forums in the USA today. Diversification in usage within and without science has made communication about "ecosystem ecology" a special semantic problem. Pickett and Cadenasso (2002) provided a thoughtful resolution to this conundrum by dividing the understanding of the ecosystem concept into three dimensions: "meaning," "model" and "metaphor." In brief, "meaning" is the broad definition of the term as articulated by Tansley (1935) without specification to a particular case or set of conditions. The "model" dimension of the term is the specification of scale, focus, and components to a particular application, usually with some abstract formulation (model) appropriate for that defined application. The "metaphorical" dimension refers to the informal, symbolic, and usually non-specific usage of multiple meanings of "ecosystems" in general parlance. In the following discussion, our usage of "ecosystem" is in terms of Pickett and Cadenasso's (2002) "model" dimension.

The structure and functions of ecosystems are typically organized by physical variables, with biological functions aggregated into those driven collectively by photosynthetic organisms, decompositional organisms, herbivorous organisms, etc. While populations are often aggregated into functional units, individual species are sometimes recognized as dominants or keys to processes within the system.

A common tenet of ecosystem ecology, stemming from its original conceptualization by Tansley (1935), is that ecosystems are systems open to inputs and outputs of energy and matter (Lindeman, 1942; Odum, 1953). Ecosystem ecologists have been quite aware of inputs and outputs although they usually emphasized vertical fluxes. In fact, graphical representations of ecosystems are typically vertical profiles designed to emphasize vertical movements. Figure 1.2, a particularly effective diagram created by H. T. Odum (1971), illustrates a commonality in the vertical structure of ecosystems ranging from deep seas with a very large vertical dimension to algal mats with very short vertical dimensions. According to this visualization, all ecosystems have an upper layer of air, an aqueous or vegetative interface with the air, a zone where

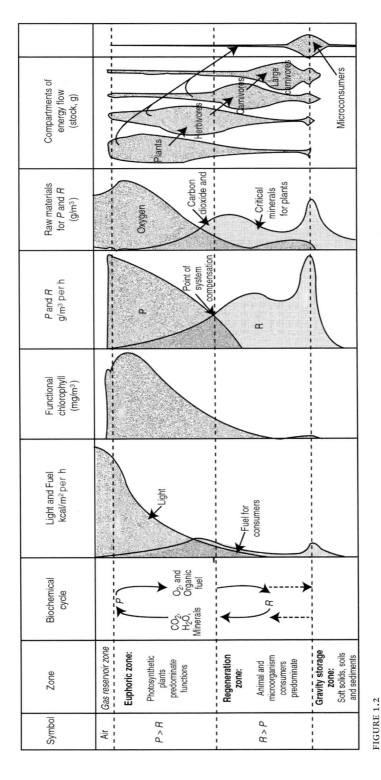

FIGURE 1.2

Vertical profiles of major ecosystem types as visualized by H. T. Odum. (From Odum, Energy, Power and Society, 1971; reprinted by permission of (Wiley-Liss, Inc., Jossey-Bass Inc., a subsidiary of) John Wiley & Sons, Inc.)

light penetrates and photosynthesis occurs, and a dark zone where hetero-trophic metabolism dominates. The companion abstraction by H. T. Odum generalizes functions and vertical fluxes for all autotrophic ecosystems in this same vertical profile visualization (Fig. 1.3). Light penetration, sedimentation, terrestrial plant uptake, and vertical circulation currents dominate this view of movements and fluxes in ecosystems. Diagrammatic models like these shape the way ecologists view nature and enhance a focus on vertical transport processes. It should be noted, however, that Odum realized the importance of lateral currents, particularly in aquatic and marine systems, by his notes at the bottom of Fig. 1.3. In general, lateral propagations relevant to ecosystem processes are now being addressed more widely at a wide range of scales in all ecosystems (Reiners, 1983; Chapin *et al.*, 2002).

Ecosystems are often viewed from three semi-independent points of view: energetics, material cycling or biogeochemistry, and regulation (Reiners, 1986). Transport processes are important in each of these. Starting with energetics, it is important to note that energy has dual components: a strictly physical component and a biological aspect. The physical energy budget of ecosystems is driven by a combination of vertical fluxes (incoming and outgoing electro-magnetic radiation, evapotranspiration, and conductive transfer to soils and sediments) and lateral or advective fluxes of sensible and latent energy.

The biological aspect of ecosystem energetics consists of photosynthesis and the transfer of organically bound energy through food webs and the resultant trophic organization. While energy transport in the physical energy budget of ecosystems is obvious, the role of transport in food webs was oversimplified in the past, as explained above. More recently, however, lateral transport has been thoroughly and convincingly reviewed by Polis *et al.* (1997, p. 308). They conclude,

> Our synthesis suggests several general principles: the movement of nutrients, detritus, prey, and consumers among habitats is ubiquitous in diverse biomes and is often a central feature of population, consumer-resource, food web and community dynamics . . . One strong insight for applied ecology is that the dynamics of seemingly distinct systems are intimately linked by spatial flow of matter and organisms.

Turning to nutrient cycling or, more broadly, biogeochemistry, movement of matter between organisms, between organisms and abiotic reservoirs, and between abiotic reservoirs is the *sine qua non* of this ecosystem viewpoint (Schlesinger, 1997). In the early 1990s, textbooks were treating plant or phyto-planktonic uptake, litter fall, leaching, and sedimentation as being mainly vertical fluxes; however, more recent literature and more applied or larger-scale treatments have placed great importance on lateral transport (Burke, 2000;

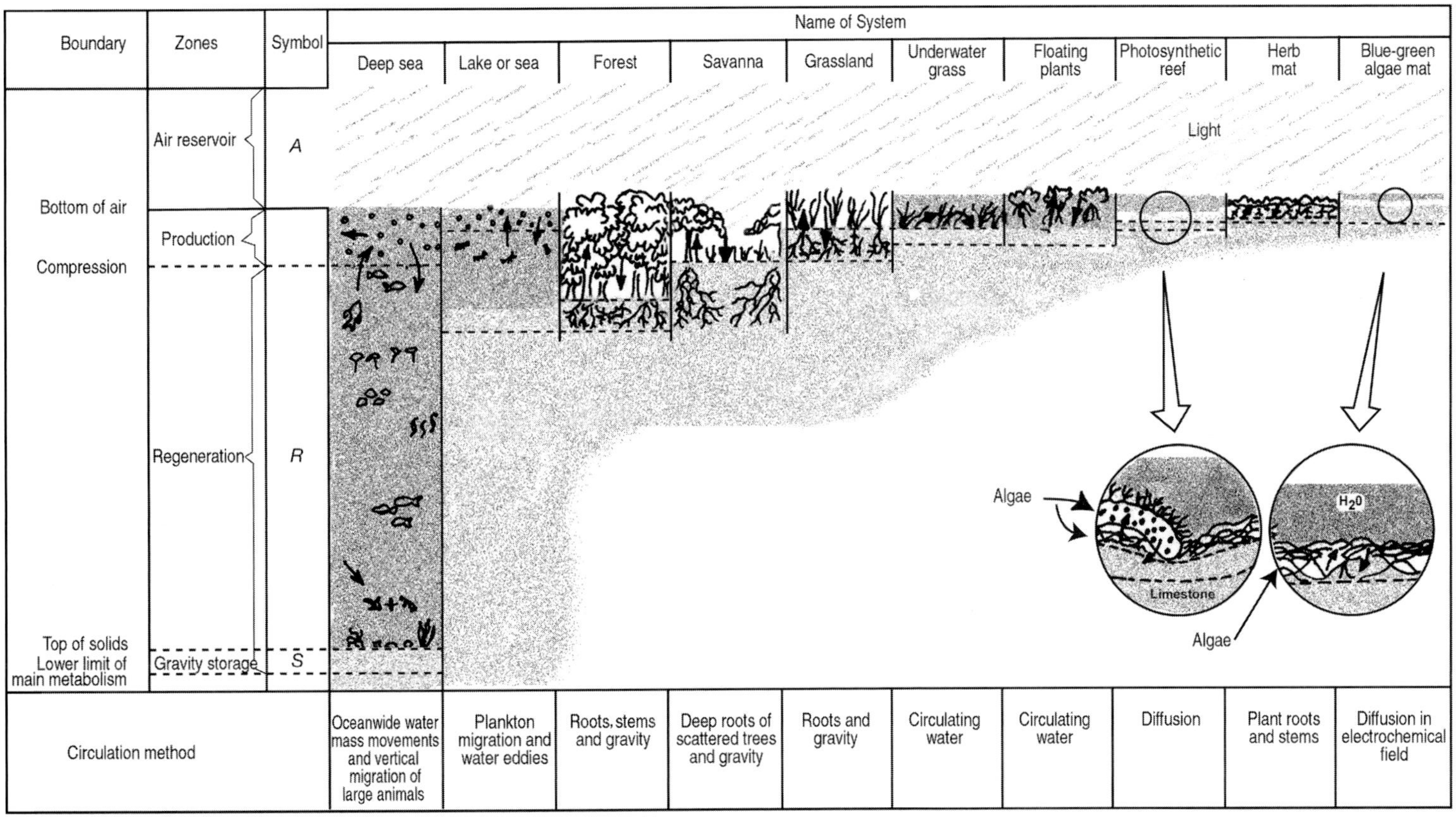

FIGURE 1.3

Generalization of autotrophic ecosystem processes in the context of vertical profile visualizations. Dominant fluxes are upward or downward. (From Odum, Energy, Power and Society 1971; reprinted by permission of (Wiley-Liss, Inc., Jossey-Bass Inc., a subsidiary of) John Wiley & Sons, Inc.).

Chapin et al., 2002). The relative importance of internal cycling versus inputs and outputs of materials depends on the boundaries set for the ecosystem in question. Generally, the smaller the scale of focus – the more localized the defined ecosystem – the more important "flow-through" is, compared with internal cycling. In addition, internal cycling is probably relatively more important in terrestrial systems than in some lakes, more in lakes than in oceanic systems, and least important in stream ecosystems. The full variety of transport processes for carbon, nitrogen, phosphorus, and sulfur, and their ranges of scales were probably first formally addressed by Reiners in 1983. More recently, Aber et al. (1999) and Meixner and Eugster (1999) recognized the importance of lateral transfers, and both Hemond and Fechner (1994) and Schnoor (1996) have described lateral transport in technical terms. Their entities of transport are all essentially matter, mainly pollutants however, and their treatments are one dimensional. That is, the transport is along one-dimensional lines or gradients rather than distributed in two-dimensional space. To a large degree, these two treatments are also restricted to ideal cases in that they do not take into account environmental heterogeneities requiring more complex transport models – a focus in later parts of this book. In contrast, Lucas et al. (1999) described a depth-averaged, two-dimensional estimation of lateral flows in a large estuary and showed how lateral movements were more important in structuring phytoplankton assemblages than were local conditions.

The science of understanding how the earth system works as a whole, often termed "earth system science," can be considered a special case of ecosystem ecology. For science at these largest extents, lateral transports are at least as important as vertical fluxes. As an example, Richey et al. (2002) have provided data to suggest that much of the carbon flux out of the Amazon Basin comes from degassing of respired materials originally drawn from the surrounding terrestrial matrix. This newly recognized CO_2 output contributes to the view of this basin as much less of a sink in the global carbon cycle.

The third ecosystem perspective – regulation – refers to those processes that lead to changes such as succession, that drive self-organizational processes such as patch dynamics in Australian rangelands (Ludwig et al., 1997), or that enhance maintenance of quasi-steady states such as gap-phase dynamics of some forests (Bormann and Likens, 1979). Regulation denotes regulating processes. What are the sources of these regulating processes? Some sources are the limitations imposed by available resources and destructive disturbances, collectively referred to as "bottom-up" control. Other regulatory mechanisms may arise from activities of members of upper trophic levels, in which case regulation is referred to as "top-down" control (Hairston et al., 1960; Pace et al., 1999). Predation, parasitism, and other interspecific interactions involved in this context are actually proximal descriptors of regulatory processes. These behaviors are derived from life history traits of constituent species. Life history traits, in turn, are ultimately derived from genetic information held within every individual of those populations. In all cases, whether top-down or bottom-up, flows and movements are involved in regulatory processes, whether they are inputs of nutrients or the dispersal of critical herbivores, predators, pests, or disease organisms (Ludwig *et al.*, 1978; Senft *et al.*, 1987; Williams *et al.*, 1996a).

1.2.5 Landscape ecology

The subject of propagation of cause and effect over space overlaps with topics and points of view reflected in the exciting and vigorous subdiscipline "landscape ecology." Landscape ecology has dual origins – human and ecological – and, consequently, markedly different traditions, definitions, and styles (Moss, 1999; Wiens, 1999; Turner *et al.*, 2001). In all traditions, however, the theme of horizontal flows and movements across landscapes has been an intrinsic part of landscape ecology and it is here, rather than in population, community, or ecosystem ecology, where the most advanced ecological treatments are generally found. Descriptions of flows and movements are at least implicitly presented in most modern treatments (e.g. Turner and Gardner, 1991a,b; Forman and Moore, 1992; Forman, 1995; Farina, 1997, 2000; Turner *et al.*, 2001); and are the explicit focus of others (Hansen and di Castri, 1992; Hansson *et al.* 1995; Rhodes *et al.*, 1996). Farina's (1997) treatment was typical of many of the earlier, general presentations on this subject. He emphasized flows of materials and movements of animals throughout his general treatment of landscape ecology, but mostly from the point of view of how discrete patterning will influence vertical fluxes, or how implicit horizontal fluxes have led to horizontal patterning. The emphasis is on how horizontal heterogeneity influences the likelihood of movements, especially of organisms, rather than on the movement process itself. Forman (1995) was more specific about how some flows, like wind, are likely to be distributed in space through physical interactions with terrain and vegetation.

Landscape ecology "deals explicitly with the causes and consequences of spatial patterns in the environment, [and] with the effects of spatial pattern on ecological processes" (Wiens, 2001a). To a considerable extent, these ecological processes are actually movements or flows of organisms, abiotic materials, and energy across landscapes as affected by spatial heterogeneity.

Wiens (1992) addressed movements and flows in a more sophisticated, conceptual framework in the context of "patches" and "ecotones" (Fig. 1.4). Patches are units in a patterned environmental space ("landscape" in this context) having some unifying quality setting them apart from neighboring units. The fundamental impact of this view of nature will be discussed in Ch. 3. Essentially,

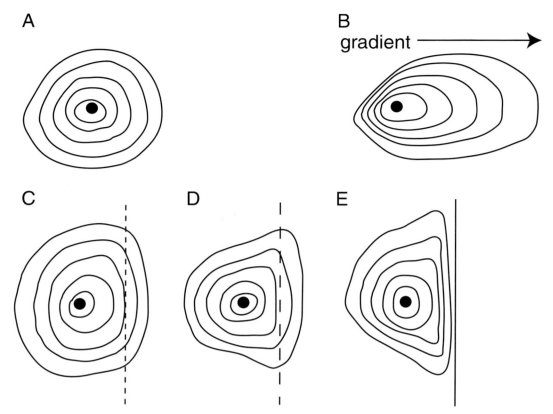

FIGURE 1.4
Hypothetical patterns of passive diffusion of materials or organisms according to
Wiens (1992). Moving entities emanate from a release point or source (dot). A is a
perfectly homogeneous environment; B is an environment with a homogeneous
gradient; C is a system bounded with a relatively permeable boundary; D is a system
with a less-permeable boundary; and (F) is a system with an impermeable boundary.

Wiens (1992) treated ecotones as boundaries between patches in a landscape
of continuously contiguous patches termed a "patch mosaic," differentiating
between unbounded systems (all one patch) and bounded systems (bounded
patches in a mosaic). Within this context, Wiens (1992) differentiated between
passive diffusion and behavioral effects. Passive diffusion is random movement
analogous to molecular diffusion: a convenient simplification of the realities
of real movement by animate or inanimate entities (Turchin, 1998). Behavioral effects apply to animal locomotion. Within passive diffusion, he noted
that there can be symmetrical movement in homogeneous patches and skewed
movement patterns in heterogeneous patches (Fig. 1.4). In this way, gradients
or other heterogeneities are built into otherwise homogeneous patches and
directionality is given to diffusional movement. With bounded systems, Wiens

(1992) discussed the probability of flows across boundaries based on either behavioral mechanisms or passive movements. He treated the former with a probabilistic approach and the latter in terms of boundaries having greater or lesser degrees of permeability to flow. Wiens' (1992) conceptual organization is very useful for general purposes and for environments that can be represented as patch mosaics.

Wiens (2001b) elaborated further on dispersal as a general phenomenon but it can be viewed as an amplification of the behavioral effects listed above. Here, Wiens illustrated how dispersal of microbes, plants, and animals is usually treated in very general terms with little regard for heterogeneity in the space over which dispersal occurs. The nature of the intervening space between point of departure and point of arrival is important to determining how dispersal occurs, how far it might occur, and how its directionality may vary in space. Landscape ecology, as a subdiscipline, should provide the means and motivation for making such evaluations.

While landscape ecologists have conceptualized flows and movements, they have, for the most part, treated them implicitly (e.g. Lepart and Debussche, 1992), abstractly (e.g. Preston and Bedford, 1988; O'Neill and Witmer, 1991; Turner *et al.*, 1993; Mowrer, 1996), mostly in one dimension (e.g. Cornet *et al.*, 1992; Ryszkowski, 1992), or, if in two dimensions, without realistic renderings of an actual piece of terrain (Turchin, 1998). General treatments, like Forman (1995) and Farina (1997), described movements and discussed their importance with only limited description of how such movements occur in mechanistic terms. Much attention is paid to corridors in these treatments, with little evidence that corridors function as postulated, a fact discussed by Turner *et al.* (2001). Movements of animals or fire, for example, are often modeled as occurring along corridors, between patches, or between patches and the matrix. These movements are most often represented as percolation or diffusion-analogue processes (not to be confused with molecular diffusion) with varying constraints across landscape units. Fluid flows of water or wind are often described as one-dimensional rather than two-dimensional processes: wind in a vertical profile and water in a horizontal profile in channels. Rarely, if ever, is any attention paid to the more diffuse saturated soil or groundwater fluxes (Forman, 1995). Colluvial flows are also treated as linear phenomena, like animal movements, that may or may not be along regular or definable networks. Swanson *et al.* (1992) provided a refreshing exception to this with their treatment of two-dimensional colluvial flows. With exceptions noted below, mechanistic representations of two-dimensional field flows (irregular, sheet-like flows across surfaces) on realistically rendered landscapes are rare in the landscape ecology literature.

Exceptions to these generalizations typically fall into one of two classes. They are either "dynamic spatial models" (also known as "process-based landscape models, e.g. Sklar and Costanza (1991)) or spatially explicit population models, which can be individual based, population based, or metapopulation based (Dunning *et al.*, 1995). Properties of these two classes of landscape model include:

1. definition of processes within explicitly represented, real-world landscapes,
2. definition of the mechanisms by which these processes take place,
3. development of means for realistically modeling such processes at the landscape level, and
4. employment of distributed movement or transport models.

Examples of both of these model classes will be illustrated in Part II.

The rarity of explicit and mechanistic treatments of flows and movements in ecology generally stands in contrast with other environmental sciences. Advances in mechanistically analyzing and modeling the propagation of entities across space are widespread in geography, hydrology, geomorphology, pathology, epidemiology, and atmospheric science. These advances typically couple mechanistic models with spatial representations like a geographical information system (GIS) as means for directly gaining understanding and making diagnoses and predictions about phenomena in particular situations (Beven and Moore, 1993; Goodchild *et al.*, 1993; NCGIA, 1996; GIS/EM4, 2000; Fotheringham and Wegener, 2000). Operational models exist but, with the exceptions noted above, are rarely found in landscape ecology literature itself. It is our thesis that explicit representation of ecological influences across landscapes can be accomplished through adoption and modification of an appropriate representation of the ecological system and an appropriate mechanistic model simulating transport. Part II will present examples of how this can be done.

Although ecology in its broader definition has always been intrinsically geographic in its nature, traditional (non-landscape) ecology has treated populations, communities, and ecosystems on a discrete site basis (Wiens, 2000). By doing so, traditional ecology outside of the landscape tradition has been conceptualized as point, or zero-order, models. The advent of comparative studies along xeric–mesic, elevational and depth, gradients brought ecology to a single axis or one-dimensional model of nature. Landscape ecology now provides the framework for two-dimensional models of nature (Turner *et al.*, 2001). Badly needed are three- and even four-dimensional models. Just as the ecosystem concept permitted the adaptation of abiotic influences into conceptualization of natural systems, landscape ecology has provided a framework for bringing spatial heterogeneity and interactions across space into our perception of nature. This is an extremely important development. While landscape ecology has conflicting language and multiple viewpoints, it has provided impetus to look at abiotic as well as biotic factors, across two- and three-dimensional space.

Despite its dual philosophical origins and multiple self-definitions (Moss, 1999; Wiens, 1999), landscape ecology is advancing ecologists' abilities to understand nature. It has already proven valuable in a number of applications, particularly cumulative impact analysis (Turner *et al.*, 2001) and conservation biology (Schumaker, 1996). Conceptualizing and modeling ecological processes in environmental space is vital to advancing ecological knowledge in general. While landscape ecology provides the conceptual context, GIS, remote sensing, and global positioning systems (GPS) provide the technical tools for making this possible. Landscape ecology and these accompanying technologies have brought us to one of the most exciting points in the history of ecology. The phenomena described in this book will be presented in the spirit of landscape ecology but are intended to expand our view from just land systems to marine and aquatic environments. We will follow our own language and definitions throughout this book.

1.3   Environmental management

Since so much of what is termed environmental management is, in fact, accommodation of extended land use and economic development, management activity usually revolves around resolution of the impacts of development on sustainable use of biological resources, maintenance of critical ecosystem functions, and assurance of species persistence. If one reviews applications of landscape ecology, ecosystem ecology, and conservation biology to actual land-management cases (e.g. Knight, 1994; Steinetz, 1996; Wiens *et al.*, 2000; Turner *et al.*, 2001), it becomes apparent that many, if not most, considerations involve the flows of materials, like water, or movements of animals or propagules. For example, shrimp production in the Bay of Fonseca, Honduras is highly compromised by sediment inundations when storms sweep across the uplands that have been converted to steepland farming. Aggravated storm flow, erosion, and sediment transport brought about by agricultural development on upper portions of the watershed have made it economically difficult, if not infeasible, to grow shrimp at the river's mouth because of the transport of sediment during storm flows (Samayoa *et al.*, 2000).

Ecology provides a general framework for policies underlying management (Turner *et al.*, 2001), but it also can and should be applied to specific cases in specific places. Ecosystem, conservation, and landscape ecology are all made more useful to management planning and execution when ecological generalities are made concrete by directly linking discrete causes with particular effects at definite geographic locations. As an example, the Wyoming State Office of

FIGURE 1.5
A drill site in a Wyoming basin. (Photograph with permission of Kerry Huller.)

the Bureau of Land Management (BLM) is currently overwhelmed with permit applications to drill for natural gas, petroleum, or coal bed methane on BLM and other lands for which BLM has jurisdiction over mineral rights. Impacts of drilling on grazing, wildlife, surface, waters, and features of historic concern such as Native American village sites must all be considered by the BLM before a permit can be granted or denied (Fig. 1.5). Under the best of circumstances, spatial data on these features are available in GIS format and can readily be made available for field office supervisors and resource specialists (e.g. Berendsen and Reiners, 1999), who can then consider the spatial relationship between the drill site together with its accompanying access road and other resources of interest. In most cases, decisions must be made by estimating how activities such as road-building and drilling influence the propagation of energy, materials, or information across the landscape surrounding the drill site. For example, drilling generates noise, which disturbs sage grouse populations. Therefore, distance to sage grouse leks is considered in terms of sound propagation, and an accurate determination of the critical distance between a sound source and the lek site is a serious ecological question. As another example, drilling with its associated road and pipeline construction often accelerates erosion downslope and into waterways and wetlands to varying degrees. The drill pad and accompanying road unquestionably impact their immediate footprint. More important is how activities at the pad and road will alter material flows and animal movements at a distance from the footprint. How far will soil move from a road cut downslope to a stream channel? Decisions must consider the transport of water and sediment in explicit terms.

Ecology in general and landscape ecology in particular are very effective at the inductive part of the scientific process: making generalizations on the basis of experiences of particular cases. Ecology is less effective, however, in the deductive mode of transferring information from generalities back to particular cases (see Turner *et al.*, 2001, p. 290). Most applied environmental problems are associated with particular places and, consequently, are unique cases. This means that better techniques are needed for deducing conclusions on specific cases from generalizations. These techniques are likely to have to include spatially specific data and consideration of propagations of ecological influences through environmental space.

1.4   The objectives of this book

This book has two main objectives. The first, addressed in Part I, is to introduce a conceptual framework for the study and understanding of propagation of ecological influences, including reciprocal interactions, across landscapes.

The second objective, addressed in Part II, is to introduce and provide examples of models that describe and predict propagation of ecological influences across landscapes. Chapters in Part II are organized by particular vectors of propagation and are designed to:

- introduce families of transport models for different transport vectors, most of which come from disciplines other than ecology;
- explain the mechanistic logic and programmatic organization of these models;
- provide interactive exercises with either a CD or interactive web site using models of heuristic value; and
- point researchers to the cutting edges of model families with citations and commentaries on the characteristics of different model sources.

Overall, our goal is to provide a systematic way to address propagation of ecological influences that will not only enhance ecology in a general and conceptual sense but will also be applicable to basic research and specific management situations.

Causes, mechanisms, and consequences of propagating influences

The thinker makes a great mistake when he asks after cause and effect. They both together make up the indivisible phenomenon.   Goethe

2.1   What do we mean by "influences" and "propagation"?

How do events or conditions occurring in one area of an environmental domain lead to transmission of effects across that domain to produce an ecological consequence somewhere else? There is no easy answer to this question. The variety of possible events or local heterogeneities is so great, and such phenomena occur at so many scales, that this subject can quickly become bewildering. The first requisite for addressing this question is a lexicon of simple, carefully defined words to describe propagation phenomena in a logically consistent manner. To begin with, there is no obvious word to convey a more general meaning of mechanistic cause and effect transmitted over a distance. We have chosen the word **influences** to convey the overarching processes linking cause and effect through space. "Influence" is the "capacity or power of persons or things to produce effects on others by intangible or indirect means" (*Webster's Dictionary*: Neufeldt and Guralnik, 1994). Admittedly, the sense of intangibility normally associated with the word is unfortunate because our goal is to be as specific as possible about the mechanisms underlying the cause and effect relationship. While influence is not an ideal word for our purposes, it does have a level of generality permitting incorporation of the many variations of influence that we address. We are deliberately avoiding the more customary terms "interrelationship" or "interaction" which occur in many definitions of ecology (e.g. Hanson, 1962) because these words specifically entail "reciprocally or mutually related" or "reciprocal action or influence" (Neufeldt and Guralnik, 1994). Influences, as we mean the term, may or may not involve reciprocity between the source of the cause and the recipient of the effect.

Initiating **causes** may be abiotic or biotic; one-way or reciprocal; rare, frequent, or continuous in time; periodic or random; localized or diffused through space. The action generated by an initiating cause usually is transferred some

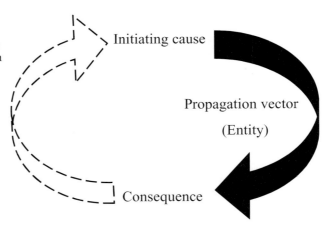

FIGURE 2.1
A simple diagram for the terminology used in this book for describing the propagation of ecological influences across environmental space. (From Reiners and Driese (2001) reproduced with permission from *BioScience* (© 2001 American Institute of Biological Sciences).)

distance, however small, to the locus of effect. Causes may be initiated by short-term events like a rainstorm or fire, or they may be more or less continuous "chronic conditions" such as the leaching of calcium-rich waters from a limestone outcrop and subsequent flow of soil solution down a slope along which the solution maintains a circumneutral pH. In all cases, some entity – dust, sound, sediment – is transported from the site of the cause to the locus of the consequence (Fig. 2.1).

We have chosen **propagation** as a neutral and general word for actual conveyance of action from its site of origin to site of effect. It is true that propagation has other meanings related to reproduction, but we are using the following meaning, "to create an effect at a distance, as by electromagnetic waves, compression waves, etc., traveling through space or a physical medium" (Neufeldt and Guralnik, 1994). "Transmission" or "conveyance" are suitable synonyms.

What is it that is being propagated? A wide variety of "things" are propagated, the definition of which depends on the effect of interest to the observer. That observer's interest dictates whether the things propagated are classified as matter, energy, or information. The general term we will use here for all of these is **entity** (Fig. 2.1).

How are entities propagated? There must be some transmitting medium or force to carry the entity from one place to another. In fact, there are quite a few carriers, ranging from random movement of molecules to the guts of large herbivores. The general term we use for all of these transmitting agents is **vector** (Fig. 2.1).

An initiating cause or condition leads to the transmission of some entity from its origin to its point of deposition at another point in the environmental domain. This deposition is likely to have an effect or several effects on components of the system at that point. Our general term for this effect of transmission is **consequence** (Fig. 2.1). The consequence may become another initiating

TABLE 2.1. *Six examples of ecological influences across landscape space defined by initiating phenomena, transmitted entities, transmission-driving vectors, and consequences of the interaction*

Initiating phenomenon on part of landscape	Entity transmitted across landscape	Vector transmitting entity	Consequences of transmission to another part of the landscape
Fire ignition	Fire front, gases, etc.	Wind	Combustion, etc.
Pathogen introduction	Dispersible spores	Wind, animals	Contamination
Precipitation events	Water flowing through or over soils	Fluvial transport driven by gravity	Erosion, nutrient transport, etc.
Snow avalanche	Snow and other debris, kinetic energy	Gravity	Destruction, redistribution of snow
Large herbivore entry	Livestock	Animal locomotion	Herbivory, differential compaction, etc.
Pollinator activation	Bees, flies, etc.	Animal locomotion, wind	Pollination

From Reiners and Driese (2001) reproduced with permission from *BioScience* (© 2001 American Institute of Biological Sciences).

cause that may lead to secondary and tertiary consequences that concatenate through the environmental domain.

As with many ecological matters, recognition of **scale** is crucial. Scale is defined here as the extent of a feature or process in space or time (Turner *et al.*, 2001). It must always be defined in absolute metrics. Undefined values such as small or large scale are entirely relative and mean different things to different people. In fact, the propagation of ecological influences occurs at spatial scales ranging from millimeters to global, and temporal scales extending from seconds to centuries.

We refrain from using the term "landscape" as a term for the environmental field of interest for two reasons. First, because our subject pertains to marine and aquatic environments as much as to terrestrial ones and the root "land" is too restrictive. The second reason is because many, but not all, landscape ecologists have definite opinions on the scale of a landscape. Some insist that landscapes are in the order of kilometers, or on the human scale, while others recognize a variable scale approach to their work (Ch. 4). To avoid an implication of a particular scale and to be inclusive with respect to environmental spheres, we use the neutral term **environmental domain** in place of "landscape."

The array of initiating causes or conditions, entities, vectors, and consequences that make up the subject field of propagating influences is vast. The rest of this chapter is devoted to a discussion of the range and variation of this array. Table 2.1 lists examples of phenomena, entities, vectors, and consequences.

2.2   Causal agents initiating the propagation of influences across landscape space

Causal agents initiating propagation of influences may be transitory phenomena (*sensu* Allen *et al.*, 1987) or chronic conditions characterizing one area of the spatial domain (Fig. 2.1). A precipitation event can lead to many consequences, ranging from wetting canopy leaves on a mountain slope to collapse of the slope itself and a debris avalanche. The entry of a fungal spore into a forest can initiate the growth of a destructive pathogen that may destroy an entire cohort of trees. A lightning strike on a ridge can lead to a brush fire that may sweep across thousands of hectares. A territorial bird-call can repel the movement of a competitor into a patch of habitat.

These examples briefly illustrate the many possible forms of causal agents. In fact, the range of qualitative and quantitative variation in causal agents is so large that a topical organization is necessary to comprehend their range and variety. Causal agents vary in seven essential ways, some of which are categorical or even binary, and some of which require continuous dimensions:

1. whether the initiating cause is a discrete event or a chronic condition;
2. the kind of environment, in the broadest terms, in which the event occurs;
3. the extent, or scale, of the initiating event or condition;
4. whether the event or condition is of abiotic or biotic origin;
5. whether the event is a natural process or an anthropogenic action;
6. the duration of the action, if it is a discrete event; and
7. the periodic or aperiodic character of the event if it is a discrete phenomenon.

Each of these means of categorization represents an axis of variation for causal agents. Although the axes differ in their length or continuity, together they represent a seven-dimensional hypervolume within which any particular event or condition may lie. These dimensions are discussed individually in the following sections.

2.2.1   Discrete events or chronic conditions?

Determining whether the initiating cause is a discrete event or a chronic condition is a question of temporal scale. A salt block placed by a rancher on a piece of rangeland is temporary in the scale of years but permanent in a scale of months. Initially, cattle attracted to the salt block will create trampling patterns radiating from it across the pasture. Those tracks will disappear in years. However, salt dissolved from the block and suffused into the soil downslope will affect soils and range plants for decades.

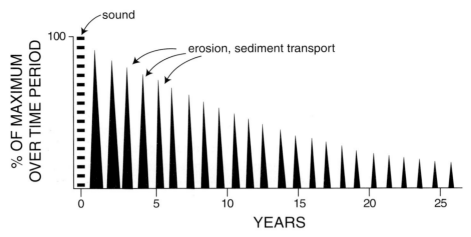

FIGURE 2.2
Projects like well installations for gas or oil initiate causes that differ qualitatively
and quantitatively in time, such as noise of installation and road erosion during the
spring snowmelt runoff over the years of pumping. Whether or not a causal event is
discrete or chronic is dependent on the time scale of the phenomenon of interest. An
individual sage grouse may be more affected by noise that we perceive, in the scale of
human lives, as short term, than by continuing sediment transport, which we
humans perceive as longer-term causation. Inasmuch as erosion only occurs during
snowmelt runoff, it is actually periodic on an annual basis, although chronic on a
10-year basis.

Initiating causes may be both discrete events and chronic conditions at different times. For example, putting in a borehole for pumping oil is a discrete
event involving road construction and scarification of a local plot of perhaps a
half hectare. But the drill pad usually will remain in place for a number of years
before any kind of restoration process is undertaken. Then, the drill pad and
associated access road become a chronic condition of that environment. Determining whether an initiating cause is discrete or chronic requires consideration
of the affected phenomenon or organism. For example, sage grouse might be
impacted by the short-term noise associated with installing a borehole, but
unaffected by the long-term influence of sediment transport (Fig. 2.2).

The relativistic nature of some of these properties is just one of many examples abounding in this and other books (e.g. Peterson and Parker, 1998) which
emphasize that analysis of the research question has to be adjusted to the appropriate spatial–temporal scale. Ecology is populated with abstract concepts such
as "species," "community," "landscape," and "connectivity," all of which must
be defined in scaling terms for effective discussion and analysis.

There is a relationship between "chronic conditions," as defined here, and
"spatial heterogeneity," as defined for landscape ecology. Some of the variations

in space that are recognized as heterogeneity and sometimes represented as patches or gradients in landscape ecology, we would view as chronic conditions for propagation processes. Wiens (2000) recognized this when addressing spatial and temporal heterogeneity in terms of function in the landscape ecology context.

> Understanding how heterogeneity affects ecological systems therefore requires an understanding not only of the environmental patterns, but also of how organisms respond to the different forms of heterogeneity (the "functional heterogeneity" of Kolasa and Rollo 1991). These responses are determined by three underlying mechanisms: movement, patch choice, and perceptual scale.
> . . . Movement is arguably the most basic of these mechanisms. Over a broad timeframe, dispersal determines how a population is structured, both demographically and genetically.

Many parallels exist between our treatment of propagation of influences in environmental space and Wiens' views on landscape ecology. One difference is the biocentrism of Wiens' organism-centered approach compared with our incorporation of abiotic phenomena along with organismal movements and actions. Usually this difference can be mediated by simply replacing the word "organism" with "phenomenon."

2.2.2 The kind of environment

The general environment in which an initiating situation occurs can be described as broadly or finely as one requires. For example, a marine ecologist can broadly subdivide marine environments between coastal zones, upwelling zones, gyres, or boundary current zones (e.g. Barber, 1988). Alternatively, a tract of marine littoral zone can be finely subdivided in terms of periods of tidal submergence. For our purposes, we will only differentiate between marine, aquatic, and terrestrial environments. This minimal level of subdivision is required because the types of causal agent, entity, and vector dominating these three kinds of environment are quite different.

2.2.3 The extent of the initiating phenomenon

The extent of the event or condition can range from millimeters to global. It is important to separate the spatial extent of the event and the domain of the environmental space over which impacts eventually may be felt. A discrete event like an earthquake may involve a crustal movement of centimeters over linear extents of kilometers, but it can affect a domain of hundreds of thousands of square kilometers. The first measure (centimeters) is the extent of the event; the second measure (hundreds of thousands of square kilometers) is the extent of the affected domain, a variable associated with the consequence of the propagation. Similarly, a chronic condition like a volcanic outcrop 20 m long can influence a downslope area of many hectares. The outcrop is the causal agent; the affected area is the measured domain in which effects occur.

The scale, or extent, of the agent that is the source of an action should be specified with accepted metrics, preferably in areal terms. Verbal descriptors mean different things to different people. The US ECOMAP program (ECOMAP, 1993) has defined ecological units on land into specified levels, possibly providing a basis for adopting general terms for areal extents larger than hectares. This program has an eight-tier, hierarchical system of units of decreasing size ranging from "domain," with a general polygon size of 2.5×10^6 km^2, to "landtype phase," with a polygon size of less than 40 ha. These peculiar units are conversions from English units still in use in the USA. Not all ecologists use the ECOMAP system in the USA (e.g. Omernik, 1987), and other countries have other systems. When working outside of a defined system like ECOMAP, it is best for users to define the extent of the system they are describing in specific areal units such as square meters, hectares, or square kilometers. Reference to "local" scale, "regional" scale, "ecosystem" scale, or "landscape" scale should be avoided, as the first two are scale independent, having no general definition, and the last two are illogical unless defined *a priori* for a particular case (Allen, 1998; Wiens, 2001b).

2.2.4 Abiotic or biotic origin? Natural versus anthropogenic action?

Whether or not an event or condition is of abiotic or biotic origin may depend on how one defines the event in ultimate versus proximate terms (*sensu* Robertson, 1989). An invasion of a fir forest by spruce-budworm moths is proximally biotic, but the arrival of the moths may, in more ultimate terms, have been facilitated by an abiotic weather event promoting their flight over long distances. The same may be true in differentiating between anthropogenic versus natural character of an event or condition. The proximal cause of a landslide may be a heavy rain on an over-steepened slope, but the ultimate cause may have been loss of root strength following clear-cut logging or slope steepening by road installation. As with other definitions and analyses, a careful recognition of scale in space and time (Wiens, 2001b) and causality in terms of proximal to ultimate phenomena is required. Ecologists must be scale aware and causation aware. The appropriate scale and best definition depend on the phenomenon being addressed.

2.2.5 Duration of discrete events

The treatment of duration and frequency can be complex. As is often the case, geographers have given such issues much thought. In her consideration of time and change, Peuquet (1999) defined change as an event or collection of events. This is consistent with the use of "causal event" in this book. She defines event as "something of significance that happens" with "significance" clearly being defined by the observer. According to Peuquet, the temporal nature of events can be classified as:

Continuous – going on throughout some interval of time
Majorative – going on most of the time
Sporadic – occurring some of the time
Unique – occurring only once.

Peuquet's taxonomy combines duration and frequency. Her "continuous" category operationally would be the same as our "chronic condition." Her "majorative" case might be regarded as chronic if the affected component of the environmental domain had a perceptual window temporally shorter than the length of time the event was occurring. If the perceptual window is longer than the duration of the event, then a majorative event becomes a discrete event. This again illustrates the centrality of the phenomenon or question for defining appropriate scales. Peuquet's "sporadic" and "unique" categories are clear examples of discrete events, but again are scale dependent.

As Peuquet (1999) pointed out, temporal patterns of event occurrence is actually another level of complexity. She said: "A temporal distribution can be chaotic, steady state, or cyclic. Similarity of states of locations or entities through time can also be converging, diverging, or combinations such as dampened oscillation." Thus, individual events can be clustered, with the clusters forming episodes; these, themselves, can be further grouped into cycles. In other words, events can be organized into hierarchical series. The temporal pattern of erosion illustrated in Fig. 2.2 is an example of periodic on an annual cycle basis but asymptotically declining over a time frame of 100 years.

As discussed above for definition of discrete event or chronic condition, definition of the duration of a discrete event is relativistic. Again, the classification of an event must be based on some metric for its duration together with identification of its effect. For example, classification of events such as inoculation by pathogenic spores as a discrete or chronic event depends upon the scale of perception. A window of 30 days may define inoculation events as discrete, whereas a window of 30 years may define the events as chronic.

Determining the appropriate level of causation in an ultimate–proximate series is also necessary. For example, a lightning strike occurring in a fraction of a second may ignite a fire, and the fire, lasting a few hours to days, may spread from

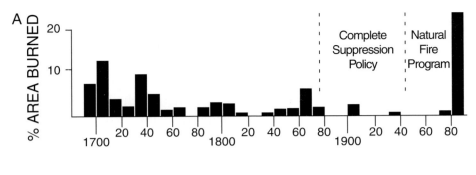

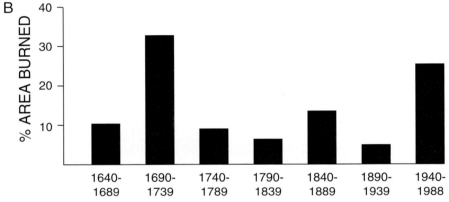

FIGURE 2.3

Extent of 129 600 ha subalpine study area on the Yellowstone Plateau burned on a decadal basis (A) and on a 50-year basis (B). Before a fire suppression policy was put into force beginning in 1886, 1–10% of the area burned every decade. Very little area was burned after that until the "natural fire program" was enacted in the 1940s, which was climaxed by the extensive 1988 fires. For phenomena like successional stage of vegetation and associated animal habitats, the premanagement period provided a sporadic range of burn events but within a relatively low bound. The period of suppression lowered this bound drastically. The natural fire program created a much more irregular situation, climaxed by the large 1988 fires. For phenomena like soil organic matter storage, having time constants of thousands of years, these variations in land burned were minor especially if viewed through the coarse resolution filter of 50-year intervals (B). (Adapted from Romme and Despain, 1989.)

its point of origin to another part of the environmental space. Is it the lightning strike or the initial fire that is the triggering event leading to a propagation of effects? If the question is about locational patterns of fire initiations, it would be the lightning strike; if the question is about fire effects themselves, it would be the fire initiation. The lightning strike is unique in a time frame of days or months but may be sporadic in a time frame of years. Similarly, fires may be unique in time frames of decades but sporadic in time frames of centuries (Fig. 2.3). If the effect in question is the timing and path of the fire, the lightning strike is unique and the better definition of cause. If the effect in question is the status of small mammal populations in the burned area, a matter involving weeks to months, the fire is unique and the better definition of the event. If the effect in question is soil-forming processes, both the lightning strike and the resulting fire are sporadic and individually irrelevant, while the pattern of occurrence of fires over centuries is important.

Even features we tend to define as permanent or chronic conditions have finite lifetimes. Our definitions tend to be scaled to our human frame of reference for space and time. An escarpment or a shale outcrop above a slope are long-term chronic conditions for phenomena measured in years, but both of these could be viewed as temporary for phenomena measured in millenia. The escarpment will eventually be reduced to a stable slope and the outcrop may eventually be buried by colluvium.

2.2.6 Periodicity or aperiodicity of discrete events

In the previous section, we observed that discrete events might occur sporadically and might be organized into repeating patterns over time. As with other definitions, the temporal pattern is dependent on the time frame of observation, which, in turn, is dependent on the phenomenon in question. It is likely that few natural phenomena are completely aperiodic if observed in an appropriate time frame. The appropriate time frame should fit the functional time frame of the phenomenon of interest. Often, this can be determined by the life span of affected organisms (White, 1979) or time scales of physical processes. Hurricanes are once-in-a-lifetime events for most birds, will be sporadic for individual forest trees, and will have an average return time for specific landscape positions within the hurricane zone (Gray et al., 1997; Boose et al., 2001). For certain geologic processes, hurricanes might finally be viewed as periodic (Murnane et al., 2000). Defining whether or not events are periodic, and, if periodic, how so, is not always such a philosophical dilemma. For most ecological issues, this will be transparent. Owl predation is diurnal; lepidopteran larval feeding is annual; but 100-year floods ostensibly occur within century-long time frames. The main objective is to define the phenomenon carefully and then choose the appropriate time frame in order to determine how the causal agent can be classified with respect to temporal pattern.

2.3   Entities propagated across environmental space

For a cause at one location to have an effect somewhere else in environmental space, there must be transmission of some entity from the source location of the cause to the location of the effect (Fig. 2.1). What is the nature of these entities? Solar flux through a canopy gap, fecal pellets drifting down through a water column, or wind-blown pollen grains wafting across a meadow are examples of such entities. These examples suggest that entities may be

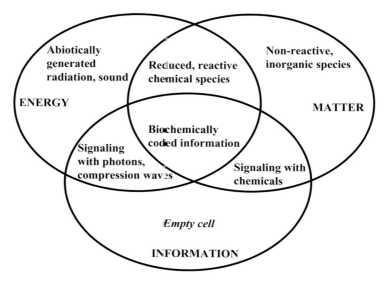

FIGURE 2.4
Entities propagated can consist of pure energy, matter, or information but almost
invariably, they are combinations of these. The essence of an entity has to be defined
in terms of its impact within the context of the phenomenon under analysis. See text
for discussion of ambiguities in defining entities. (From Reiners and Driese (2001)
reproduced with permission from *BioScience* (© 2001 American Institute of
Biological Sciences).)

characterized as being parcels of energy, matter, or information that are transmitted across environmental space (Fig. 2.4).

Closer consideration shows, however, that most entities are combinations of
these. For example, it is difficult to imagine any entity that might not, depending on the consequence, have some information value. Abiotically generated
light and sound are listed as examples of entities in the pure energy cell of
Fig. 2.4, while non-reactive inorganic materials like plagioclase crystals are
given as examples in the pure matter cell. However, under the right circumstances, each of these might be interpreted as bearing information as well.
Sunlight may contain spectral or daylength information for organisms capable
of sensing it, and soil organisms can determine the difference between calcium
and potassium ions in the soil solution. Kangas (1989) would define all entities
as manifestations of energy. Upon careful reflection, it becomes evident that
we cannot say *a priori* that any entity would be pure energy or pure matter.
Membership of any entity in one of the cells in Fig. 2.4 is contingent on the
eventual effect of that entity in that particular case and, therefore, can only be
defined *post hoc*, after the effect is defined. Again, the relativism of variables in
propagation to the specific phenomenon becomes apparent.

While entities defined as energy and matter may take on alternative values, information is unambiguous. There can be no entity consisting of pure information. Information, whether behavioral signals or genetic code, must be encoded in either energetic terms, such as sound waves, or in material terms, such as pheromones or genetically dispersable materials like spores. Consequently, the pure information cell in Fig. 2.4 is an empty cell. An entity bearing information must always contain energy, matter, or both. Nevertheless, information is a valid and extremely interesting kind of entity.

In one fundamental way, all material entities possess the energy of motion – momentum – if they are propagated through space. When momentum of the entity is, itself, the effect of interest, then the entity must be put into the energy class.

2.3.1 Propagation of energetic entities

Better known forms of propagated energy are kinetic (momentum of moving mass), sensible heat (thermal energy of mass measured as temperature), latent energy (energy in the physical state of water), chemical energy (bound energy of reduced substances such as organic matter), and electromagnetic radiation. A log rolling down a hill possesses both physical momentum and chemical energy. In the short term, the energy of momentum leads to an initial consequence as the energy of momentum is dissipated into other objects, but the chemical energy bound in the log also becomes available to decomposing organisms at this new locale, thus leading to a secondary, delayed consequence. Microclimate, at the scale of centimeters to meters, and much of our weather, at the scale of thousands of meters, depend on transport processes. Both are driven by radiation transport and by advection flows of sensible heat and the energy represented by water vapor (latent energy). We also can suffer damage from, or derive useable energy from, the momentum of the atmosphere. Thus, a parcel of air can transmit energy in three ways: as momentum, sensible heat, or latent heat. This is in addition to possible energy content of reduced compounds in the air such as nitrous oxide, methane, or carbon monoxide.

Electromagnetic radiation and sound waves are significant kinds of propagation with a number of similarities. Electromagnetic radiation consists of photons behaving as though they are waves of particular wavelength and intensity. Electromagnetic radiation ranges from very short, highly energetic cosmic rays to ultraviolet wavelengths, through the visible spectrum into near infrared to far infrared and out into the long wavelengths of radiowaves. Most examples of ecologically relevant electromagnetic radiation transmission involve wavelengths ranging from ultraviolet, seen by some insects and vertebrates, through visible light (defined by human retinal capacities) to near infrared, or even thermal infrared radiation (Goldsmith, 1990; Bradbury and Vehrencamp, 1998).

Sound waves are compression waves created by the compression of molecules in a local area of the physical medium: air, water, soil, and rocks. Ultimately, this is a special form of kinetic energy. Individual molecules move outward and then rebound, creating concentric waves of compressed molecules propagating away from the source. While matter is involved in these compressive waves, there is no net movement of molecules composing the medium, just a local oscillation (Bradbury and Vehrencamp, 1998).

A less-common form of energetic propagation is the generation of bioelectric fields and reception of electrical signals. Bradbury and Vehrencamp (1998) described how certain fish species living in murky waters communicate with variable electrical discharge rates, even modulating their discharges in ways that seem analogous to "songs" by birds. While a fascinating and valid example of informational propagation, this phenomenon will not be pursued further here.

Electromagnetic radiation, sound, and electroreceptivity are unusual examples of propagated entities in that no net material transport is associated with the energy. This makes such propagation the nearest to a pure signal in the usual sense of the term. Readers are advised that animal behaviorists, having argued long and hard about signals in that context, have specific definitions and criteria for signals and cues, although these definitions are not consistently and universally observed (Bradbury and Vehrenkamp, 1998). Radiation and sound can have other effects unrelated to animal communication; for example, thunder can trigger debris avalanches, a propagation process ultimately involving considerable mass.

2.3.2 Propagation of material entities

Excepting electromagnetic radiation, sound, and electrical fields, most examples of propagation involve transmission of matter, whether it is gases, liquids, solids, or dissolved and suspended solids in water. Water running down slopes and channels and through soil and aquifers incorporates the mass of all of these material forms. Colluvial movement of hillslope materials or snow involves mostly solid matter but may include substantial amounts of water. Particulates, including spores and nutrients, gases, and the mass of air itself, are transported by wind. Enormous masses of water, together with organisms, particulate, and dissolved substances, are moved long distances in marine environments by tidal and non-tidal currents. Minuscule amounts of mass are moved by pollinating insects. Water, gases, and fine particulates are usually transported over distances of millimeters to centimeters through fluid media by molecular diffusion.

Fire is a special case in which energy and matter are mixed in complex ways. At a flame front, several entities are involved: the radiant and conducted heat of the flame, the mass of hot gases, and the kinetic energy of moving air. Once again, entity definition is contingent on the effect of interest.

As already noted, any mass movement involves both the translocated mass as an entity, and the kinetic energy associated with the momentum of the mass. Sometimes it is the kinetic energy of the mass rather than the material itself that is more important as an entity.

2.3.3 Propagation of information entities

The third category of entity – information – is itself complex through the diversities of meaning that information may take. We use the common language definition of information here as "knowledge communicated or received concerning a particular fact or circumstance" (Neufeldt and Guralnik, 1994) realizing that even this definition leads to semantic issues about "knowledge."

If we divide information into two categories, communications and genetic information, we encounter yet another level of semantic confusion. The definitions of communications are as complex as the definitions previously reviewed for events and entities. For an authoritative discussion of communication as animal behaviorists use the term, readers are advised to review a source such as Bradbury and Vehrencamp (1998). For our purposes, we define "communication" as any state or event that alters, or potentially alters, the state or behavior of a receiver. This definition includes information generated by abiotic phenomena such as wind as well as the more specific cues, signals, and highly evolved communications described by animal behaviorists. In our usage, the information source might be abiotic or biotic, but the receiver would have to at least potentially be organismal. We would not include the sound of thunder in triggering a landslide as a propagation of information, for example. That simply would be propagation of energy, which vibrates the slope mass sufficiently to trigger its movement. The term "potentially," used with respect to an organismal receptor, signifies that the transmission of an information-bearing entity occurs whether or not a percipient receiver actually senses the transmission and becomes influenced by it. By this reasoning, a flash of color radiating from a male bird display is considered the propagation of information whether or not a female of that species sees that flash. Likewise the diffusion or wind-transmitted molecules of a pheromone would be propagation of information whether or not a target receiver was even in the vicinity. This definition is at odds with usages of others but serves our purposes regarding propagation of influences.

The second category of information is genetic information. It is the DNA or RNA codes carried by spores, pollen, seeds, other vegetative propagating material, as well as eggs and dispersing animals. Such information is encoded in material form combined with associated free energy, but when the principal interest is in the transference of genetic material, propagating entities like seeds are viewed as information. We acknowledge that an ambiguity exists here as to whether an entity can be information if it is not decoded by a receiving element of the ecosystem. Consistent with our inclusion of "potential communication" discussed above, we adopt the view that, like an unopened newspaper, encoded information is still information whether it is decoded or not.

2.4   Propagation vectors

Entities must be propagated by vectors. "Vector" has a mathematical definition as a quantity possessing both magnitude and direction. Vector also has a biological definition, usually used for organisms that transmit pathogens. We are generalizing vector to mean any agent providing transmission of an entity across space. Synonyms might be "carrier" or "conveyor," although they too have other nuances. Wiens (1992) used vector in this same sense in the landscape ecology context.

Definition of vector agents requires choosing an appropriate operational level in the logical sequence of ultimate to proximal causation. Ultimately, all vector agents operating in Earth's domain are derived from four sources: nuclear phenomena, gravity, the momentum of Earth's rotation, and living organisms. Nuclear reactions lie behind solar radiation and tectonic movements; gravity lies behind most other movements, and the rotation of the Earth directs large-scale movements of the atmosphere and oceanic waters. A relatively small amount of energy, originally derived from solar radiation, is the ultimate source of animal locomotion except where food chains are based on inorganically reduced compounds, as in marine rift zones.

While philosophically interesting, these ultimate sources of causation are not useful to our discussion. What is needed are more operational categories that fall between the ultimate causes and an exhaustive range of special examples. We have selected 11 basic vectors operational on Earth that are medial between ultimate causes like solar radiation and highly proximal causes like a particular kind of slope movement (Table 2.2). These 11 vectors are molecular diffusion, fluvial transport, colluvial transport, glacial transport, sedimentation, tidal currents, non-tidal currents, wind, sound, electromagnetic radiation, and animal locomotion. Each of these can be subdivided into increasingly

TABLE 2.2. *Medial transport vectors responsible for transport of ecological influences within the Earth's domain together with their ultimate causal agents*

Medial causal agent	Ultimate causal agent(s)
Diffusion	Solar radiation, radiogenic heating
Fluvial transport	Solar radiation, gravity
Colluvial transport	Radiogenic heating, gravity
Glacial transport	Solar radiation, gravity
Sedimentation	Solar radiation, gravity
Tidal currents	Gravity
Non-tidal currents	Solar radiation, Earth rotation, gravity
Wind	Solar radiation, Earth rotation, gravity
Electromagnetic radiation	Solar radiation (directly and through animals)
Sound	Miscellaneous
Animal locomotion	Solar radiation

detailed manifestations of the general vector. For example, fluvial transport can be subdivided into surface and groundwater, between saturated flow and unsaturated flow, etc. Both electromagnetic radiation (light and radiant heat) and sound stand apart from other transport vectors in that the vector is also the entity propagated.

2.4.1 Diffusion

Diffusion is the transfer of mass by the movement of molecules or very small particles because of their kinetic energy (Monteith and Unsworth, 1995). The amount of kinetic energy is a function of temperature of the material. Random movements caused by collisions between molecules lead to random distributions, finally resulting in complete and even particle dispersion within the constraints of the enclosing environment. The gas laws learned in secondary school chemistry courses are the perfect mental model for diffusion.

Diffusion is enormously important over the vast interface surfaces that make up the Earth. These include interfaces of sediments with water, soils with the atmosphere, vegetation with the atmosphere, and the ocean with the atmosphere. Diffusion is ultimately responsible for much of the gas and water transport that goes on between organisms and their surrounding media, whether through integuments like skin, or specialized organs like gills. While diffusion takes place over enormous surface areas, the distance of diffusive movement is usually very short. Diffusion typically is superseded by some form of mass flow over longer distances than millimeters to centimeters, depending on the physical state of the medium. For example, water vapor diffuses out of leaf mesophyll to stomatal openings and into the surface boundary layer surrounding leaves. From there, however, it is transported into the surrounding air by turbulent transport processes involving small-scale thermal eddies or larger-scale atmospheric turbulence. As another example, pheromones are produced by many animals and diffuse outward until they are caught up in mass flow processes. It is mass flow that brings pheromones to a hive so that a swarm of African bees can find the perpetrator who kills one or is stung by one. The practical applications of diffusive transport are explored further in Ch. 6.

The term diffusion is also used by analogy for fluid transport at particular scales (Chs. 7 and 10) and for the postulated or apparent random movement of organisms (Ch. 11). In both cases, this analogy is a useful modeling convention for cases where it is not useful or possible to model movement deterministically. For animal movements, diffusion, as an analogy, is basic to percolation models: movement models incorporating specific rules involving the connectivity between nearest neighbor sites or sites having common edges (Gardner and O'Neill, 1991; Wiens, 1992; Turchin, 1998). Inasmuch as this book is dedicated to examples of explicitly mechanistic transport, when we describe diffusion as a fundamental vector, we are restricting our usage to the sense used in physics, that of atoms, ions, molecules, or small particles moving as a function of their kinetic energy. Other uses of diffusion as an analogy will be made clear as they occur.

2.4.2 Gravity-driven vectors

Gravity is the ultimate driving force for several of the transport vectors listed in Table 2.2. Newtonian physics describes gravity but does not explain it. A fundamental explanation is still sought but current theories require general relativity and, more recently, string theory (Greene, 1999). This book is constrained to phenomena at or near Earth's surface so that the traditional Newtonian interpretation is adequate and the term "force" is appropriate. Functionally, gravity is a force field causing fluids and solids to move downward if they are situated beyond a critical angle of repose. Thus, gravity underlies transport by running water, mass wasting, glaciers, tides, and sedimentation in fluids. It also plays a role in atmospheric, lake, and oceanic movements.

While considering transport by gravity-driven vectors, it is appropriate to consider the vectors that move material in opposition to gravity. Endogenic tectonic movements, driven by mantle heating processes of mostly radiogenic origin, drive crustal materials above base level. This book will not treat tectonics as a transport mechanism because of its irrelevance for ecology in the short term. In the long term, however, crustal tectonics is highly relevant to ecology through its creation of continents, islands, and submarine features on Earth, as well as the maintenance of altitudinal gradients along which gravity-driven vectors can operate. Similarly, solar radiation operates in opposition to gravity by heating air and evaporating water. Moisture-bearing warm air rises, sinks, and flows laterally in response to Coreolis effects as planetary winds. These processes underlie the global transport of sensible heat and water vapor over enormous distances, and the deposition of snow or rain to higher altitudinal positions upon which gravity can act. On a global scale, and in appropriate time frames, transport by tectonics, solar heating of air, evaporation of water, and gravity underlie most, but not all, of the energy and material fluxes that sustain ecological systems. In a sense, this book is about smaller-scale propagation phenomena operating within the environmental flux fields at the global scale. These global-scale flux fields are established and maintained by the interplay of radiogenically driven tectonics, solar radiation, and gravity.

2.4.3 Colluvial transport

Colluvial transport is used by us synonymously with the geomorphological term "mass movement," defined as "the downslope movement of slope material under the influence of the gravitational force of the material itself and without the assistance of moving water, ice or air" (Summerfield, 1991). Geomorphologists have identified many kinds of colluvial transport depending on the mechanisms, materials, moisture content, nature of movement, and rate of movement. Summerfield (1991) has presented a table of 24 types, including rock creep, debris avalanche, snow avalanche, soil creep, rock fall, and settlement. We usually relate colluvial transport to areas of steep terrain, but solifluction movements can occur across surprisingly shallow slopes under wet and freezing conditions. Colluvial transport obviously involves the movement of mass, but rapid mass movement events such as rock slides or mud flows also involve huge amounts of kinetic energy focused on relatively small areas over brief spans of time.

2.4.4 Fluvial transport

Fluvial transport is a general term for movement of water and incorporated materials driven usually by gravity but sometimes by winds (Summerfield, 1991; Julien, 1999). This, too, is a complex subject decomposable into many different processes that can be subdivided even further. On land, fluvial transport ranges from small-scale process like rain splash erosion to the multiple routes of subsurface flow, to overland flow to channel flow. Fluvial transport is equally important below ground. Groundwater transport has become an environmental and scientific issue of enormous consequence throughout the developed world (Schnoor, 1996). In lakes, estuaries, and seas, currents play enormous transport roles, but they are treated later in this chapter as separate categories from fluvial transport on land.

Fluvial transport is an immense and sophisticated area of research underlying hydrology, geomorphology, limnology, marine science, atmospheric science, engineering, and other aspects of the environmental sciences. The common theme in our usage is the role of liquid water moving in response to gravity as a conveyance mechanism. Polis *et al.* (1997) presented a thorough review of fluvial transport of matter and energy (as nutrients and detritus) in terms of water to water, land to water and water to land. We will return to fluvial transport in Ch. 10.

2.4.5 Glacial transport

Glacial transport is the transport of solid and liquid water with their incorporated solid materials through the erosional and depositional mechanisms of ice movement. The driving force is, of course, gravity. Glacial transport is contemporaneously important only in regions of mountain and continental glaciers that still exist today (about 10% of continents). However, glacial transport was enormously important over large areas of the northern hemisphere during the Pleistocene and Holocene Epochs (1.6 million years ago to the present), when glaciers covered about 30% of the continents (Summerfield, 1991), and was a dominant force during the Proterozoic when all land and most of the Earth's surface were covered with ice. Glacial transport has been the dominant cause of contemporary landforms over the vast, recently deglaciated surfaces of the world. Because glacial erosion is so powerful and rapid, it is the primary determinant of how high mountains are likely to be on Earth (Broecker, 1986; Brozovic *et al.*, 1997). Enormous amounts of material ranging from clay to very large boulders can be transported in moving ice. Most glacial debris is inorganic, although some detrital organic matter is also imbedded in ice to be released and deposited, sometimes with sensational results as when long-deceased mastodons or even human corpses make their appearance in the melting ice. This book will not examine glacial transport further but more information can be found in Summerfield (1991) or Eyles (1983).

2.4.6 Tidal transport

Tidal movement is a powerful transport vector in the marine coastal zone. Tides are generated by the gravitational force of the sun and other planets, but especially the moon. In fact, the combination of forces involved are fairly complex, but a simple explanation is that sea level bulges occur on the side of Earth closest to the moon and on the opposite side of the Earth where centrifugal force from Earth's spin combines to produce a minimal gravitational force. Depressions in seawater level occur in positions away from these areas of bulge (Neshyba, 1987). The resulting differences in sea level vary at any one place on Earth's surface in a number of ways, but typically there are one or two periods of high water level (high tide) and low water level (low tide) per day. Because of differences in the shapes of sea bottoms and coastlines, some areas have tides of many meters whereas other areas have scarcely measurable tides (Levinton, 1995). The movement of water up and down leads to strong horizontal displacements of large amounts of water and, thus, lateral tidal currents in the coastal zone.

Tidal currents tend to be higher in areas of greater tidal amplitude and in semi-enclosed portions of coasts such as estuaries. From a geomorphological point of view, tides are particularly important for moving unconsolidated sediment and less important for erosion of compact or indurated material (Summerfield, 1991). Laterally moving currents rather than waves are the characteristic form of tidal transport in estuaries (Levinton, 1995). From a biological point of view, tides dominate the physical environment in the intertidal zone, although tides are involved in deep ocean mixing as well (Pinkel *et al.*, 2000). The periodic submergence and emergence of these zones with attendant contrasts in environmental resources and stresses control the zonation and organisms occurring in this rich and productive environment. The movement of water, usually as waves superimposed on currents, along open coastlines is the primary form of transport for water, nutrients, organic detritus, and organisms. These materials pass across sections of the estuary, creating the strong temporal dynamics in community structure and ecosystem dynamics that are of such fascination to ecologists (Polis *et al.*, 1997). For further information on tides, see Neshyba (1987).

2.4.7 Sedimentation

Sedimentation is the downward flux of materials in fluids in response to gravity. These fluids include the atmosphere, lakes and rivers, oceans, and magma chambers. Very often, sedimentation is combined with lateral movements associated with winds and currents; however, by definition, gravity remains the underlying force driving these sedimentation fluxes. Sedimentation is one of several means by which particles arrive on Earth from space (Johnson, 2001), leave the stratosphere for the troposphere (Tabazadeh *et al.*, 2001), and are transported from the troposphere to be deposited on land and sea surfaces. It is difficult to differentiate sedimentation from other atmospheric depositional processes like rainout, diffusion, and impaction, which are more closely related to motions of air. Nevertheless, sedimentation is a major source of particulate deposition (Hicks and Matt, 1987).

Sedimentation in lakes and oceans involves both inorganic and organic matter. Sedimenting material may be recirculated and resuspended; once it reaches lake or sea floors, it may undergo chemical transformations (diagenesis) and eventually be converted to sedimentary rocks (lithification). From a biological and biogeochemical viewpoint, it is the sedimentation of organic matter with its incorporated nutrients that is of greatest interest. Incoming mineral detrital material, marl (calcium carbonate), and planktonic debris sediment out of the epilimnion of lakes to the hypolimnion. Depending on the turnover properties of the lake, this material may be permanently trapped in the sediments, or the organic portions may be decomposed and mineralized and returned to the photic zone when and if lake turnover occurs (Wetzel, 1983). The same process is equally important in oceans, although considerable spatial variation occurs in terms of zones of high productivity and sedimentation versus zones of nutrient regeneration (Redfield, 1958; Lenton and Watson, 2000). In the oceans, lateral transport by currents between sedimentation zones and mineralization zones, and consequent return to the photic zones, may involve tens of thousands of kilometers and transport times in the order of hundreds of years, depending on the ocean basin (Murray, 1992).

2.4.8 Non-tidal currents

Non-tidal movements by lakes and ocean currents are driven by planetary winds, Earth's rotation, and gravity. Lakes have a broad range of possible movements and currents that are usually unique to each lake (Wetzel, 1983). One of the most important movements in many lakes are those associated with the alternate development and destruction of thermal stratification, often referred to as lake turnover. This can happen in many ways. Shallow lakes can turn over continually; some lakes may only turn over annually, occasionally; or never (Wetzel, 1983). The classic condition, often taught in ecology courses, is for lakes to turn over and mix twice a year (dimictic stratification). During periods of thermal stratification, there is little to no mixing between the upper depths and lower depths of the lake. When stratification is broken down by changing weather, storms, or winds, mixing can resume between the layers. Turnover and other water movements in lakes are vital to the regeneration of nutrients and maintenance of primary productivity and consumer metabolism.

Oceanic surface currents are set in motion on the surface by the planetary wind system and are directionally modified by the Coreolis effect and interactions with continental margins (Levinton, 1995). These surface currents are matched by deep-water circulation that is driven by gravitational effects on water of different densities, by the Coreolis effect, and by boundary constraints set by continents. Differing densities result from differences in temperature or salinity, so the overall circulation system varies geographically with heating, evaporation, cooling, and freshwater inputs into regional waters (Neshyba, 1987). Deep circulation, usually referred to as "deep thermohaline circulation," is linked with surface currents to generate a large-scale current system possessing generalizable geographic patterns, velocities, and water turnover times. Readers can refer to oceanographic references for details of this circulation phenomenon. To a considerable degree, the ecological conditions of a point anywhere in the three-dimensional volumes of the oceans is a function of the relationship of that point with respect to circulation systems bearing nutrients to that point (e.g. Williams and McLaren, 2000) or exporting organic matter from that point (e.g. Laws *et al.*, 2000).

Oceanic circulation also accounts for transport of materials of global significance such as trace gases (Suntharalingam *et al.*, 2000), major atmospheric components such as oxygen and carbon dioxide (Archer *et al.*, 2000; Sarmiento *et al.*, 2000), and heat. In fact, the global marine circulation system, referred to as the "thermohaline conveyor," is considered to be a central control on atmospheric chemistry and the global climate of Earth (Broecker, 1987, 1997). In spite of the enormous mass involved in this circulation system, it has a surprisingly short turnover time. Within oceanic basins, it takes only two to three years for surface water to circulate around its gyre (Neshyba, 1987). Deepwater turnover times vary for different oceanic basins, but range from 250 years in the Indian Ocean to 510 years in the Pacific (Murray, 1992). This may be considered the fundamental circulation system that maintains a buffered global environment and sets constraints on smaller-scale regions and ecosystems in the time scale of hundreds of years (or less, according to Broecker (1997)). On shorter time scales, in the order of years to decades, variations in marine currents are known to have potent influences on weather. Changes in circulation patterns associated with the El Niño–southern oscillation of the Pacific, with time periods of two to ten years, are widely recognized to have powerful influences on marine ecosystems and continental weather (Ely *et al.*, 1993; Murray *et al.*, 1994). This book does not further explore aquatic or marine currents as transport vectors but readers are invited to see Neshyba (1987) for a treatment of this topic.

2.4.9 Wind

Excepting gravity and diffusion, wind is the most ubiquitous transport vector experienced by organisms living at the surface of Earth. At the planetary level, wind is generated the rotation of the earth, by unequal distribution of incoming sunlight and heating, and by rising, cooling, and sinking of air. In the simplest terms, greater solar radiation around the equator heats air, causing it to rise and move poleward. As it moves poleward, it is deflected to the east by the Coreolis force, producing a general westerly wind direction. Return flow from high latitudes to the equatorial regions is similarly deflected to the west, producing general easterly winds. For more details, see Green (1999). Wind at the planetary level is central to climate and responsible for mixing the gases and particles that are entrained in wind. Wind below the planetary boundary layer (1–2 km) becomes shaped by local forces of eddy formation at all scales, from kilometers to millimeters, and by interactions with topography. Wind close to the ground and in contact with surface boundary layers is also controlled by eddies and local heating and cooling cells. The treatment of wind ranges from the planetary scale (e.g. Green, 1999) to the planetary boundary layer (Stull, 1988; Garratt, 1992), to the surface boundary layer (Monteith and Unsworth, 1995; Raupach, 1995), with each segment in this continuum representing appropriate scales of wind function.

Obviously, wind transports atmospheric gases and entrained particulates. It also transports sensible and latent heat and water vapor, all elements of energy and water budgets for ecosystems at every scale. Besides mass and energy transport, wind is also a major vector for propagation of small animals, spores, pollen, and other small disseminules: forms of biological information (Isard and Gage, 2001). Entire ecosystems exist in otherwise inhospitable environments like caves, mountain tops, snowfields, and polar regions where the energetic and nutrient resources are derived entirely from transport and deposition by wind (Polis et al., 1997). Wind provides critically limiting nutrients to oligotrophic portions of the open ocean (Jickells et al., 1998; Fung et al., 2000), and even to infertile land areas (Chadwick et al., 1999) through the deposition of dust. Wind is responsible for global distribution of anthropogenically produced nitrogen, leading to many potential changes in regional ecosystem function (Matson et al., 1999). The mechanisms of wind transport are further reviewed in the application presented in Ch. 8.

2.4.10 Electromagnetic radiation and sound

Propagation by both electromagnetic radiation (light and radiant heat) and sound were discussed as entities. Both cases stand apart from other transport vectors in that the vector itself is also the entity propagated. Applications of these two vectors to ecological phenomena are presented in Chs. 12 and 13.

2.4.11 Animal locomotion

Locomotory capacity can be found in members of all biological king-doms (regardless of how kingdoms are defined) in some stage of their life cycle, but for the most part, this topic refers to animals and what some call protists. By their own locomotion, whether crawling, swimming, running, or flying, they transport themselves and distribute their effects of feeding, deposition of waste materials, trampling, pollination, etc. across environmental space. In addition to moving themselves and translocating their direct effects, animals can be propagation vectors for other organisms such as parasites and plant disseminules. Humans are the ultimate animal transport vector in all of these ways.

Some animals may be only weakly locomotory in an environment of powerful transport flows, like larvae of coastal sessile forms (Cowen *et al.*, 2000). Other animals, such as Arctic terns, which migrate from one circumpolar region to the other, have extraordinary locomotory capabilities. To the extent that animals move autonomously, they are an independent vector and are not driven by other forces like gravity. The diversity of animals is such that they follow no single set of rules and the generalization of their movement is very difficult. Nevertheless, modeling animal movement has received significant attention because of the intrinsic interest and importance of animals. Turchin (1998) has presented a superb synopsis of general modeling of animal movement. Dunning *et al.* (1995) reviewed the state of the art of deterministic movement modeling for animals. An example of this vector will be presented in Ch. 11.

2.5   Consequences of propagation of ecological influences

Propagation of entities by vectors leads to consequences somewhere in environmental space through the sequence of steps outlined in Fig. 2.1. The primary consequences may be the only and final results, but some consequences have the potential for begetting secondary consequences at the same point in space over time. On the one hand, a rock may roll down a hill and come to rest, causing no measurable immediate consequence whatever. On the other hand, the rock may roll down the hill, knock over a tree, open the canopy, create a tip-up mound, initiate a population explosion of tree seedlings, and create a lair for some animal. In this way, a series of secondary consequences would result over time.

Another possibility is that a consequence can become a new triggering event, which will lead to the subsequent propagation of an influence to another locus in the environmental domain. Such subsequent propagations may be entirely

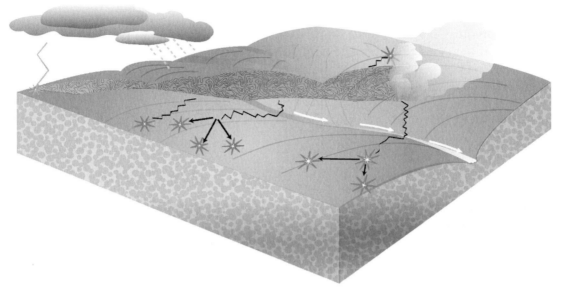

FIGURE 2.5
Schematic diagram of the concatenation of possible influences initiated by a single cause across an environmental domain. Lightning at a single point initiates a fire, which is carried by strong winds across the domain. Smoke, bearing water vapor, soot, ash, and other gases, is carried across and out of the domain by wind. Dissolved and suspended materials are eventually carried out through the drainage flowpaths of the domain by fluvial transport. Colluvial transport along the steeper drainage ways may be accelerated by the fire. Animals fleeing the fire will alter their movement paths which may have consequences for them, their prey, and their predators.

different in their properties from the initial propagation, involving different entities and vectors. When an initial event initiates a chain of propagating phenomena, it creates a concatenation of cause and effect over space in the environmental domain (Fig. 2.5).

2.6   Propagations and surface forms

Many of the transport processes discussed earlier in this chapter have an impact on surface features of land, the coastal zone and, to a more limited extent, the ocean floor. They can do this either through long attrition or by episodic events. Glacial transport reshapes valleys originally shaped by fluvial transport. Changes in wind direction redistribute sand dunes. Tidal currents control the reaches, depths, and courses of estuarine channels. The waves and currents generated by large storms create new bars or destroy old ones. In these and many more ways, both individual propagation events and more-or-less continuous transport processes shape Earth's surfaces. These are the exogenous processes operating in opposition to endogenous processes that control Earth's surface features.

There is a reciprocity to this relationship between transport processes and surface features. Surface features also influence propagation processes. All other things being equal, drainage along slopes is likely to be faster on terrain with greater relief than where there is lesser relief, and it is more likely to follow entrenched channels than to move along planar surfaces. This is so obvious to geomorphologists that surface features are visual cues to which transport processes are of greatest importance in a given environment (Swanson *et al.*, 1992). Swanson *et al.* (1987a) elucidated this by recognizing four classes of landform effect on ecosystem patterns and processes.

> Class 1: Landforms – by their elevation, aspect . . . , parent materials, and steepness of slope – influence air and ground temperature and the quantities of moisture, nutrients, and other materials . . . available at sites within a landscape.
> Class 2: Landforms affect the flow of organisms, propagules . . . , energy, and material . . . through a landscape.
> Class 3: Landforms may influence the frequency and spatial pattern of nongeomorphically induced disturbance by agents such as fire, wind, and grazing.
> Class 4: Landforms constrain the spatial pattern and rate or frequency of geomorphic processes that alter biotic features and processes.

Of course, the importance of these classes is dependent on climate and, particularly, relief (Moore and Burch, 1986a). Solifluction is not a factor in warm climates, and snow avalanches are not issues in flat terrain. However, the extent of topographic variation required to make an ecological difference seems to decrease in proportion to the relief available. For example, in gently rolling to nearly flat terrain, very small changes in topography or proximity to the capillary fringe of water tables can make large differences in ecological conditions for plants.

The same points made for landforms by Swanson *et al.* (1987a) are made by coastal zone scientists (e.g. Ray and Hayden, 1992) and to a lesser extent by deep-water marine scientists. The basic structure of the physical environment, whether terrestrial, aquatic, or marine, will set constraints on the kinds and degrees of physically driven propagation and will, at the same time, indicate which vectors are likely to be operating in that environmental domain.

2.7   Propagations and disturbance

It is easy to appreciate the consequences of propagating events and conditions from the point of view of ecosystem disturbance. White and Pickett (1985) defined disturbances as discrete events in time that modify landscapes, ecosystems, communities, and population structure by changing the substrate, the physical environment, or the availability of resources. Certainly, the phenomena described here include disturbances as so described. Slope failure in steep, mountain terrain leading to massive colluvial transport to the foot slope followed by sediment delivery into an adjacent stream and consequent destruction of benthic populations is an example of concatenated propagation events that would, in another context, be termed disturbance. However, not all propagation phenomena fulfill any definition of disturbance. A frog's mating call, sent across a marsh, is certainly not a disturbance but is an important ecological influence propagated across environmental space.

Some "disturbance" syndromes themselves move across space in predictable patterns. Several of these were described by A. S. Watt in his 1947 seminal paper *Pattern and Process in the Plant Community*. Watt's insight into endogenously generated cycles of local disturbance and regeneration has, since that time, caused ecologists to perceive that pattern in vegetation or marine littoral communities is a clue to process, and that processes may generate a stable or unstable mosaic of structural features. These structural features (patches), in turn, may have characteristic functional dynamics. Since Watt's insight was published, ecologists have described other self-organizing patterns of death and regeneration that may be all, or in part, endogenously driven. A particularly dramatic example, because it involves trees, are "waves" of death and regeneration by trees in subalpine zones of New England and Japan (Sprugel, 1976; Kimura, 1982). Matson and Waring (1984) described a similar phenomenon driven by different processes in mountain hemlock forests of Oregon. Ludwig *et al.* (1997) described dramatic patterns in Australian semiarid rangelands in which bare ground surrounds elongated bands of trees and other plants. These bands are oriented perpendicularly to the slope direction and are products of "run-off–run-on" flow of water and organic matter across the slope. These patches tend to migrate upslope because net resource gains from above foster a vegetational growth front upslope, and because net losses of water, nutrients, and organic material from fringes at the lower margin of the band leads to plant dieback.

These cases suggest that disturbance itself propagates across environmental space. At one level of interpretation that is a reasonable view. Our interpretation is, however, that some influence is propagated across the landscape which, operating in conjunction with endogenous processes, drives the generation of repeating patterns in ecosystem structure and function. It is the momentum of wind, or the drying capacity of wind, or the propagation of fungal spores by hyphae along gradients of plant vigor, or the flow of water down slope that are the operational agencies propagating over space, not the disturbance itself. The disturbance or, conversely, constructional processes result from propagation of definable entities underlying the phenomena of change.

2.8   Propagations as ecological linkages through space

A generally accepted definition of ecology is the study of interrelationships of organisms with one another and with the environment (Hanson, 1962). For the purposes of this discussion, we relax the part of the definition demanding reciprocity between interrelating elements and include all influences, both one-way and reciprocal. Interrelationships can be, and are, studied individually. However, it has been recognized historically that ecology has a need to organize interrelationships collectively in order to make generalizations about them. Such organizations have been promulgated in a number of ways. One traditional way has been organization of constituent species into systems of competition/predator–prey linkages and evaluating them in demographic terms (e.g. Holling, 1959). Another way, practiced by behavioral ecologists, is to record interactions within an animal community in terms of the cues and signals between individuals (e.g. MacArthur, 1958). Still another approach, related to the predator–prey demographic approach but associated with ecosystem ecology, has been to quantify energy flow through food chains (e.g. Teal, 1962). Yet another approach has been to delineate feeding patterns (Brooks and Dodson, 1965), trophic groups (e.g. Hairston *et al.*, 1960), guilds (e.g. Root, 1967), or critical community roles such as keystone species (Paine, 1966).

All of these approaches to organizing and generalizing about interrelationships have merit and suit different needs. We suggest here that the cross-space propagation system we have presented in this chapter is another way of organizing the complexity of interrelationships in ecological systems. If we consider the wide variety of propagation phenomena occurring at all spatial and temporal scales, it becomes evident that environmental space is traversed by untold numbers of propagations, each with their own causal events, entities, vectors, and consequences. Figure 2.5 is a hypothetical, but not unrealistic, example of how several kinds of event, not all occurring simultaneously or even in the same time frame, can create interactions having lasting effects across an environmental domain.

At this point it is important to consider time. Propagations are transmitted over varying distances, cover varying areas or volumes, and also occur over varying periods of time. Consideration of time brings in consideration of past along with present events or conditions. Propagations that occurred in the past can leave a consequence as important as impacts from ongoing events. The passage of a fire or the dispersal of pathogens can leave traces in the environmental domain that last centuries to millenia. The ghosts of propagations past have to be considered along with those of the present in this complex network of interactions.

The combined effects of past and present propagations have possibly always been understood by at least some ecologists as part of the "interactions" that underlie the mystery and intractability of ecological systems. Admittedly, not all so-called interactions necessarily involve transmission of influence over space. When a leopard pounces, a mosquito injects a plasmodium, or bower birds mate, some kind of transmission takes place over space, but the travel distance involved compared with the scale of the actors makes such interactions local in nature. We have no basis for concrete definition of when a local action becomes a propagation across space. Tentatively, propagation over a distance of ten times the length of the agent or area occupied by a chronic condition might qualify such an action as a propagation. However, we are reluctant to erect hard rules for all phenomena. Discrete rules probably are best left for particular cases when operational definitions are mandatory.

2.9   A general model for space–time events

In this chapter we have elaborated on variations in initiating events, entities, and vectors to illustrate the range of settings, scales, and forms that propagation of ecological influences can have. A sensitivity to propagations makes it possible to view nature in terms of flows and movements, propagated causes and effects, rather than, or as well as, a collection of physical objects. The range and variation of these phenomena are very large. Needed is a system for generalizing all of these phenomena in common terms. This subsection describes a general model to fulfill this need.

Propagation events are intrinsically spatial–temporal phenomena. Propagations move across space over time. Propagation time may be short in human terms, as with light and sound; other times, it can seem slow by human reckoning, as with slope creep. Rates depend on the transmitting vector and the distance traveled by a discrete entity like a spore; the area involved by a distributed entity like wind will depend on the phenomenon and the environmental situation. The point is that propagations are both space and time phenomena. A good general model should reflect this.

The relationships of space and time for these phenomena are highly situational. How can we generalize about this relationship to induce some general rules for this class of phenomena? In his appropriately titled paper, *Process dynamics, temporal extent, and causal propagation as the basis for linking space and time,* Kelmelis (1998) provided a general analysis of the relationship between space and time for process modeling in spatial–temporal contexts. He defined "process" as "changes that lead toward a particular result" and correctly pointed out that time is implicit in change. According to his argument, "geographic processes space (S) and time (T) are inextricably linked through the dynamics of the processes (D_p), and the communication of effects of the process, that is, the causal propagation (CP)." Therefore, S and T are related through some function (*f*) of (D_p, CP). The spatial extent of the area impacted by the process (E_{sp}) is a function of the process D_p such that

$$E_{sp} = f(D_p). \tag{2.1}$$

Kelmelis acknowledged that the relationship of this function varies widely among processes. In general, spatial extents resulting from propagation events are positively related to the time since their initiation. That is, spatial extent will increase over time since the event took place. However, the spatial range of each propagation event depends on the magnitude of the event (M), the temporal extent of the causal event (E_t), and environmental factors influencing the propagation rate. For example, the spread of a pathogen will generally increase over time, but the exact rate and extent will depend on weather and dispersal conditions. Similarly, a snow avalanche will cover more ground with time, but its full "run-out" distance will depend on slope steepness and vegetative resistance to flow. These limiting environmental factors are defined by Kelmelis (1998) as resistance or viscosity to causal propagation through a medium (V_{cp}) and to attenuation processes (A).

Kelmelis (1998) pointed out that there can be secondary and tertiary effects of events. For example, he illustrated that hurricanes have primary effects of wind and rain, secondary effects of floods, and tertiary effects of societal dislocations and economic loss. These secondary and tertiary effects are the same as the "concatenating effects" discussed above. The duration time of the event (E_t) is the time from initiation to completion of one or all levels of effects depending on how one wishes to define them.

Kelmelis (1998) defined the causal propagation function as:

$$C_p = f(t_e, V_{cp}, A) \tag{2.2}$$

where t_e is the elapsed time since the onset of the communicable portion of the event; V_{cp} is velocity of causal propagation, which is, in turn, a function of resistance to causal propagation of the medium through which the event is communicated, and A is an attenuation factor.

The transformation is,

$$E_s = f(D_p, CP) \tag{2.3}$$

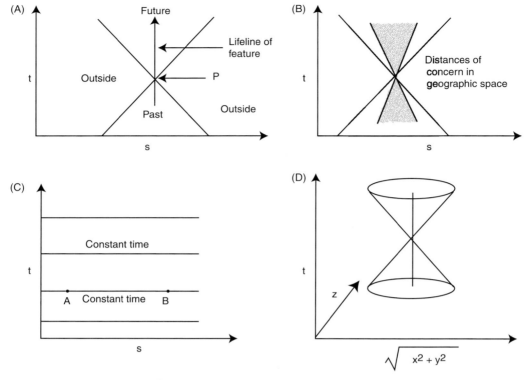

FIGURE 2.6
Relativistic space–time (A, B) and classical space–time (C, D) for the distribution of
a propagated event. (From Kelmelis (1998) in *Spatial and Temporal Reasoning in
Geographic Information Systems*, edited by Max Edenhofer and Reginald Golledge,
copyright 1998 by Oxford University Press, Inc. Used by permission of Oxford
University Press, Inc.)

and by substitution,

$$D_p = f(M, E_t) \quad \text{and} \quad CP = f(t_e, V_{cp}, A) \tag{2.4}$$

resulting in,

$$E_s = f(M, E_t, t_e, V_{cp}, A). \tag{2.5}$$

Consequently, spatial extent is related to the magnitude and temporal extent of
the event, elapsed time since the event was initiated, and the velocity modified
by attenuation factors of propagation of the event.

Kelmelis (1998) generalized the space–time relationship as a diffusion process. As emphasized earlier, not all events are propagated as molecular diffusion
or even as analogies of diffusion. However, diffusion is useful as a general and
illustrative case. The space–time "track" of the propagation process is illustrated as a two-dimensional "diffusion cone" over time in Fig. 2.6a where

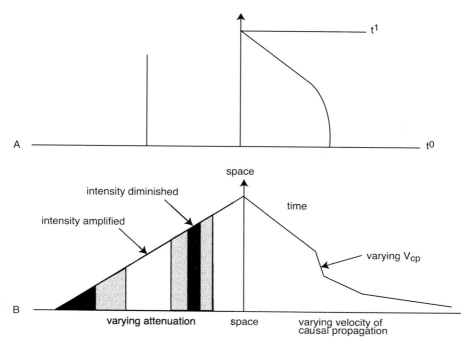

FIGURE 2.7
The effects of attenuation and velocity of propagation transmission. (From Kelmelis
(1998) in *Spatial and Temporal Reasoning in Geographic Information Systems*, edited by
Max Egenhofer and Reginald Golledge, copyright 1998 by Oxford University Press,
Inc. Used by permission of Oxford University Press, Inc.)

current existence of a propagation event has a past and a future. The cone
boundaries are limited by velocity and attenuation factors. The two different two-dimensional boundaries in Fig. 2.6b illustrate the extent of affected
zones depending on velocity and attenuation of the causal propagation process.
Figure 2.6c is a Newtonian representation of all space at one moment where the
horizontal lines are time slices. Events A and B appear to occur simultaneously
in space in one time slice. Figure 2.6d is a representation of an event occurring
in three-dimensional space and one-dimensional time.

The effects of attenuation and variable velocity are illustrated in Fig. 2.7.
In Fig. 2.7a there is complete attenuation on the left space axis of the path of
influence in time and space, and a uniform reduction in velocity of transmission
on the right space axis. In Fig. 2.7b, differential velocities result in the irregular
path boundary on the right side of the axis whereas diminution or amplification
can affect the path boundary as shown on the left axis.

Figure 2.8 illustrates how secondary effects can lead to another diffusion
cone, represented here as a two-dimensional path. In this representation,
Kelmelis (1998) showed a case in which attenuation may be high but an

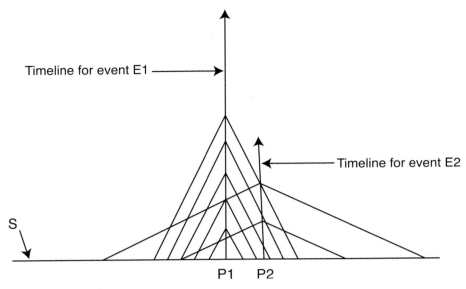

FIGURE 2.8
The effect of different velocities of communication for different events on a surface. Event E1 occurs at point P1 and affects the surface (S). At point P2, event E2 is caused by the effect of E1. This takes place at time *t*. The transmissivity of the surface to the effect of E2 is greater than to the effect of E1; therefore, the spatial extent of the effect of E2 rapidly increases beyond that of E1. (From Kelmelis (1998) in *Spatial and Temporal Reasoning in Geographic Information Systems*, edited by Max Egenhofer and Reginald Golledge, copyright 1998 by Oxford University Press, Inc. Used by permission of Oxford University Press, Inc.)

extended elapse time can extend beyond the path boundary of the first propagation set by attenuation limits.

Kelmelis provided a valuable general model for the relationship between space and time for propagation events. The cone (in three-dimensional space) or path (in two-dimensional space) of influence is a product of the elapsed time since the event and is modified by the magnitude of the event, the length of time the event generates an influence, and environmental factors affecting propagation rate. Each of these factors can be translated into appropriate physical terms for different vectors, events, media, and environmental space to fit particular cases. These cases will differ from the idealized diffusion behavior and homogeneous environment used for building a general case. Examples used later in Part II of this book will show how different vectors will interact with explicit environmental space to give specific direction and distribution to the entities propagated.

How do we see nature?

There is no final ecological truth. All knowledge is a current approximation, and each addition to that knowledge is but a small, incremental step toward understanding.   Jack Ward Thomas *Wildlife in Old-growth Forests*, 1992

3.1   The issues

In order to address the topic of this book – how propagation of ecological influences occur in environmental space – it is necessary to establish a meaning of the term environmental space as we use it in this book. Careful discussion reveals that psychological and philosophical disparities exist in how we both perceive and conceptualize nature. Like "ecosystem," the term nature means different things to different people. Those of us in the natural sciences recognize the cognitive differences between natural scientists and "the others" about the nature of nature. Ongoing political debates in the USA about evolution versus creationism, and the origin and age of the Earth, highlight these disparities. Less obvious to natural scientists are the interesting and significant differences existing in this regard between different kinds of environmental scientist and even between members of a special disciplinary subset – ecologists. This chapter is dedicated to the elucidation of these differences and concludes with a conceptualization of nature necessary for the representations of environmental space to come in Part II of this book.

3.2   Individual cognition of nature

In everyday life, we take for granted that what we see, hear, smell, feel, and taste around us is reality. Casual conversation about common sensory input seems to confirm the view that we all hear and interpret the same things in the same way. Two strollers hearing a robin's evening song remark to one another that it is pleasant to hear robins sing in spring. However, deeper discussions between the strollers about the robin's singing may turn up some surprising differences in their understanding of what they heard. One stroller may take the poetic viewpoint that the bird is singing in spontaneous exhilaration; this

FIGURE 3.1
View of burned terrain in Yellowstone National Park in 1989. For some, this is this a scene of disaster, for others an all-too-infrequent natural event. (Photograph by William A. Reiners.)

listener would vicariously share the bird's happiness in the scene. In contrast, the other stroller may have heard a more aggressive intent in the bird's song. He or she may have been imagining territorial establishment and maintenance and may have been enjoying their witnessing of adaptive behavior in action. While pleasant emotions may be felt by both, their understandings of the phenomenon they each heard are quite different.

Cognition of nature goes beyond the direct sensory perception and involves more abstract conceptualization. For example, when Yellowstone National Park experienced widespread forest fires in 1988, many considered this a disaster (Fig. 3.1). For them, the Park had largely been destroyed. These people (including prominent politicians) pilloried the National Park Service for "permitting" these fires to run their course. At the same time, others took pleasure in witnessing natural processes becoming manifest in spite of fire control efforts; in this case, attainment of episodic events with long repeat times. For this latter population, this was an historic event equivalent to putting men on the moon and they rejoiced in anticipation of an ecologically healthier pattern of meadow and forest patchworks in the Park.

The Yellowstone fires of 1988 were a historical event that not only marked the Yellowstone landscape for centuries but also altered the attitudes of observers in terms of how they saw nature. Because of the repercussions of this event toward subsequent management policy-shaping, it is an archetypical dilemma of matching perceptions with realities in order to make policy decisions. Whyte (1985) defined environmental perception in these terms.

> Environmental perception is the means by which we seek to understand environmental phenomena in order to arrive at a better use of environmental resources and a more effective response to environmental hazards. The processes by which we arrive at these decisions include direct experience of the environment (through the senses of taste, touch, sight, hearing and smell) and indirect information from other people, science, and the mass media. They are mediated by our own personalities, values, roles and attitudes. The study of environmental perception has to encompass all these means of processing environmental information and to place the individual psychological processes of prediction, evaluation and explanation into a relevant social and political framework.

Whyte provided data on the incredible range of perception individuals have on environmental conditions and hazards, which highlighted the variation among people in this regard even when issues involve personal well-being such as their own health, and when the data are sound and readily available.

Shepard (1991) has shown how, within the western European tradition, perceptions were strongly shaped by the attitudes toward nature and ways in which perceptions had been trained by mythos and art prevailing at the time. A mountainous landscape could be considered a frightening wasteland in one historical period, and a sublime vision in another.

Perceptions can change even within the lifetimes of individuals, for better, or for worse. William Wordsworth (Todd, 1996) sadly recognized the melancholic fact that even single individuals may lose some abilities to "see nature" with the following lines from *Ode: Intimations of Immortality*.

> There was time when meadow, grove, and stream,
> The earth, and every common sight,
> To me did seem
> Apparelled in celestial light.
> The glory and the freshness of a dream,
> It is not now as it has been of yore; –
> Turn wheresoe'er I may,
> By night or day,
> The things which I have seen I now can see no more.

Even among members of presumed identical educational–cultural–experiential classes, cognition can vary in startling ways. Two plant ecologists viewing a pine savanna in mountain foothills may perceive the same visual image but report quite different cognizance of what they see. One may "see" vegetational expression of an environmental gradient such as moisture availability while the other may "see" the product of differential fire frequency. While these differences in how we see nature may be interesting sociological and psychological phenomena in themselves, they are addressed here because they pose linguistic and philosophical obstacles to communicating about the propagation of ecological influences.

3.3   Geographic conceptualizations of space and time

Propagation of influences through environmental space is about time-varying processes occurring in space. Space and time are domains of geographers, who have much to offer in this regard. Sophisticated analyses of fundamental aspects of spatial and temporal reasoning by humans illustrate the complexity of this subject (Egenhofer and Golledge, 1998). Even within the sphere of more formalized treatments of time and space, there are multiple points of view (e.g. Mark, 1999; Martin, 1999; Pickles, 1999; Raper, 1999). Here, we follow the presentation of Couclelis (1999), an exposition particularly amenable to our environmental perspectives. According to Couclelis, systematic thinking in the modern intellectual tradition about time and space has its roots in four disciplines: geography, mathematics, physics, and philosophy. These root areas overlap with one another, as illustrated in Fig. 3.2. Geography is the most ancient of the ways to treat space and time, necessitated by the requirements of humans to explore, exploit, and defend space. It originally was an empirical approach that became codified in maps, charts, and tables.

In order to improve descriptions and explanations of spatial–temporal problems, people turned to mathematics. Geometry evolved as the mathematical language of space; later, calculus emerged as the mathematical language of time. Space and time were usually represented in Euclidean terms in which definite metrics are defined and quantifiable. An alternative viewpoint of relative space and time, associated with Leibniz, permitted space to be defined as the set of all possible relations between phenomena. A familiar example would be the relationships between two trains moving in opposite directions on parallel tracks, defining a shrinking, then vanishing, then again expanding space between them. In relative space, time can be defined analogously as the order of succession of phenomena. More recently, mathematics has developed methods for treating topological space. Topology involves the invariant properties of figures or objects such as inside and outside, right and left, touching and

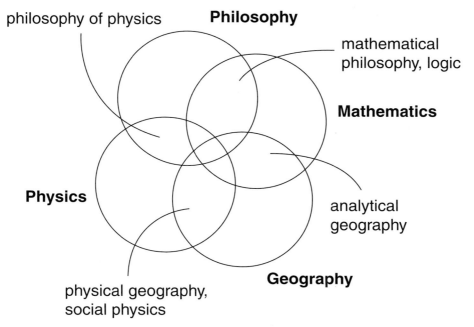

philosophy of physics

Philosophy

mathematical
philosophy, logic

Mathematics

Physics

analytical
geography

physical geography,
social physics

Geography

FIGURE 3.2
The historical roots of spatial–temporal perspectives. (From Couclelis, H. (1999)
Space, time, geography. In *Geographical Information Systems*, 2nd edn, ed. P. Longley,
M. F. Goodchild, D. J. Maguire, and D. W. Rhind, pp. 29–38. New York: Wiley.
Copyright © 1999 John Wiley & Sons, Inc. Reprinted with permission of John
Wiley & Sons, Inc.)

overlapping, being connected with, and connectivity itself. This fundamental spatial relatedness is obviously very important in environmental science in
general and to GIS in particular.

According to Couclelis (1999), physics' contributions to views of space and
time were made in close conjunction with those of mathematics. Newton's
mechanics were set in Euclidean space and time (absolute space) in contrast with
Leibniz's relative space. The Newtonian view dominated for several centuries,
but in the twentieth century physics turned to relative space and time to address
certain classes of problem.

Philosophers raise other types of issue such as the perennial question of
whether space and time are objective properties of the world or just constructs
of human understanding. Philosophers also debate whether the primary ingre-
dients of the world are "things" or "properties." Adoption of the first view leads
to an ontology of objects; adoption of the second leads to an ontology of fields.
"Ontology" may be even more disparate in its meanings than is ecosystem. Even
among philosophers, there are multiple points of view. According to *Webster's
Dictionary* (Neufeldt and Guralnik, 1994), ontology is that branch of metaphysics
that studies the nature of existence as such, as distinct from material existence,

spiritual existence, etc. Ontology is a popular term in geography pertaining, in this discussion, to the meaning of time and space. According to geographers, different ways of thinking lead to different ontological viewpoints of space, time, and spatial–temporal systems. Raper (1999) provided an exposition of the ideal natural scientist's perspective, or ontology – an ideal from which we sometimes depart, as will be discussed below in the ecological context.

Couclelis (1999, p. 33) pointed out that, in our postmodern time, two other perspectives have joined these traditional four in providing perspectives on spatial–temporal phenomena: the cognitive and the sociocultural. She says,

> Both are based on the premise that there is no single objective reality that is the same for all, but that different realities exist for different minds or for different sociocultural identities. This implies that the world as described by mathematics and physics is not the only world there is, and that in fact the world so described may be of little relevance to people's thinking and activities.

The cognitive realm of spatial–temporal conceptualization is about how individuals view and deal with the world. This serious subject can be illustrated by the humorous and insightful example by Steinberg published on the cover of *The New Yorker* magazine in 1976, illustrating how a native New Yorker views the world (Fig. 3.3). For the Manhattanite living on 9th Avenue, New York City looms large. New Jersey and Pennsylvania are extensive neighboring regions, but the remainder of the country is compressed into a vague, "large-ish" area east of San Francisco, while the rest of the world is barely seen on the horizon. Steinberg had the topology right but the metrics were highly distorted – correctly, in all likelihood – to represent a Manhattanite's cognition of the world. This cartoon has been reprinted many times, providing non-New Yorkers an opportunity to make fun of New York chauvinism while giving New Yorkers an opportunity to show how little they think of the rest of the world. For some of us, though, Steinberg's representation goes deeper than that as we realize that we all see the world in much the same way from wherever we live. For those of us living in the US Central Rocky Mountains, the 1100 miles to Chicago is somehow further than the 1100 miles to San Francisco. Our ontologies of space and time are very much distorted by individualistic cognitive maps.

The second postmodernist perspective – the sociocultural perspective – involves ontologies held in common by particular groups. One example involves differences between how Native-Americans and Euro-Americans see the same geography. The former include properties of sacredness, creation, and community. Such attributes are not, and possibly cannot, be included in conventional geographic representations of Euro-Americans (Pickles, 1999). We suggest that different kinds of environmental scientist, ranging from oceanographers to

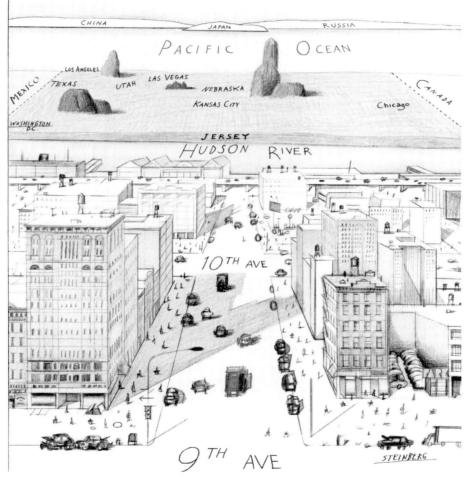

FIGURE 3.3
"View of the World from 9th Avenue" by Paul Steinberg. (Copyright © 1976 by *The New Yorker* Magazine, Inc. Reprinted by permission of the Estate of Saul Steinberg/Artists Rights Society (ARS), New York, and *The New Yorker* magazine, Inc.)

epidemiologists to ecologists of various stripes, represent sociocultural groups with their own characteristic ontologies of space, time, and nature.

3.4   Ontological perspectives on time and space in the environmental sciences

The title of this chapter – How do we see nature? – refers to everyone, but in our context pertains primarily to environmental scientists in general and ecologists in particular. "Seeing" is a bit ambiguous, involving at least three levels of understanding nature. "Perception" properly refers to mental sensations and processes emanating from sensory systems and occurring directly from sensory stimuli (Mark, 1999). Such stimuli are, themselves, subject to chemical/optical and neurological distortion. "Cognition" refers to conscious thinking, including memory and reasoning along with perception (Mark, 1999). Beyond this, there is much debate about the nature and kinds of cognition. As we go through our lives, we probably move between a kind of "experiential" cognition that is more or less an automatic registration of stimuli into a preformed conceptual structure, and a more abstract form of cognition in which stimuli are reworked into more complex concepts and models. Certainly, expressing cognition in text, maps, or digital representations involves translation into abstract symbols. From this, it would seem possible that an ornithologist hearing the robin's song may first put that perception into a territoriality frame of reference in his or her experiential mode. Later, the ornithologist may also have reason to map graphically or digitally the location of *the* singing robin for the purpose of calculating the shape and extent of that individual's territory. These would be acts requiring operation in a more complex cognitive mode involving the use of abstract symbols and models.

Environmental scientists, by which we mean scientists addressing earth or biological processes, such as hydrologists, oceanographers, geomorphologists, atmospheric chemists, pedologists, and ecologists, have some common points of departure for conceptualizing nature. They probably have the same general cosmological beliefs and the same understanding of Earth's geography and the physical processes dominating the crust, oceans, and atmosphere. Nevertheless, in many ways, members of the various environmental disciplines operate in different worlds. The phenomena addressed, and the kinds of environment in which environmental scientists work, erect major divisions in conceptualization and approach. Understanding, modeling, and predicting hydrodynamical flows in the coastal zone, or fluvial erosion by small streams, or chemical reaction rates in the stratosphere will naturally cause individuals to see nature quite differently from those who dissect the contents of fish guts to determine the allometric relationship between predator fish and their prey.

As well as phenomena themselves, the kinds of environment in which scientists work further modify ontological views of nature. Scientists working in the physical domains of atmospheric parcels, seas, lakes, streams, or on land have differing ontological perspectives of how nature is arrayed in space. If one queries environmental scientists as to their understandings and personal meanings of "nature," "the natural environment," and the nature of a particular place and time, quite different responses are encountered. Part of this variability results from whether or not the subject is responding as an individual/citizen or in their professional role. For some such individuals, terms like "nature" are only abstractions; for others they evoke imagery. Images that come to their minds usually include lakes, streams, mountainous vistas, desert scenes, or rocky seashores. Often, these images are related to the environment in which they live or work. Ask a pelagic zone marine biologist about an abstract scientific idea and he or she is most likely to place that idea into marine environments dominated by fluid surfaces, volumetric dimensions, and arbitrary lateral boundaries. Put a similar question to a soil scientist and he or she is likely to think in terms of continuous variation in three dimensions of a semisolid with arbitrarily defined limits. Similarly, some vegetation ecologists (see discussion below on ecological traditions) may put the idea into the framework of a continuous volume of varying thickness and composition, a volume enclosing plant shoots and, less likely, roots. An animal ecologist, terrestrial or aquatic, might be more likely to envision a volume including objects – usually animals in a matrix of other objects, like plants. A paleoecologist may adopt an idea similarly to other ecologists, but they also may project a surprisingly vivid image of ideas regarding an earlier time frame.

Individuals having different cognitive frameworks for new sensory inputs will handle such inputs differently. Some will focus on surfaces, some on volumes. Those scientists dealing with discrete entities, like animals, may conceptualize nature as a place where objects occur in volumes co-populated by other objects. Those focusing on continuous fields may mentally organize spatial heterogeneity as continuous variation, as in a fine-grained raster, as a mosaic of discontinuous spatial units (patches), or as continuous vector space.

Environmental scientists dealing with continuous, physical properties, such as terrain, atmospheric and water waves, sound, or light, over extensive domains may envision mathematically defined, repeating, spatial–temporal patterns. Especially among oceanographic and atmospheric scientists, phenomena can be represented mathematically and graphically – and apparently mentally visualized – as characteristic wave patterns. Interacting phenomena can be seen to have their wave patterns of characteristic frequency and amplitude, or even nested sets of wavelets. Scales of these waves can be in meters or seconds and up to thousands of kilometers or months (cf. Kantha and

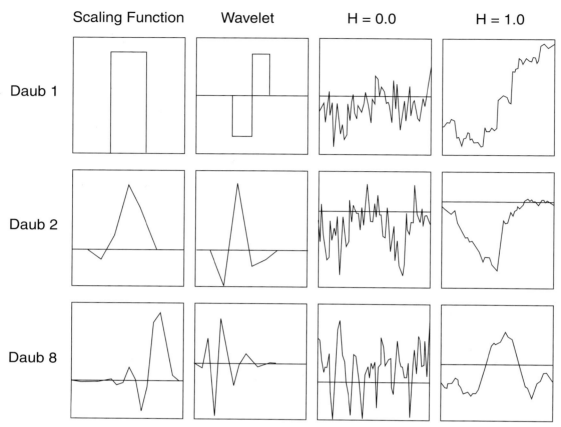

FIGURE 3.4

An explanatory schematic for wavelet synthesis of one-dimensional transects of a property of a landscape. "Wavelet bases are Daubechies (1992) series of order 1, 2, and 8. Wavelet order determines the texture of the pattern: lower-order wavelets produce sharper edges; higher-order wavelets are more continuous." (From Keitt (2000). Reproduced with permission of the author and Kluwer Academic Publishers.)

Clayson, 2000a,b). This way of seeing nature is apparently less common among terrestrial scientists although there are known applications (Bradshaw and Spies, 1992; Keitt, 2000) (Fig. 3.4).

Conceptualization of time depends on the temporal scale of the phenomenon with which individual scientists most often work, whether it is the characteristic time scale of waves, local wind eddies, diurnal physiology, annual migrations, biennial lake turnover, semidecadal El Nino–southern oscillation events, or millennial-scale soil leaching. In this regard, environmental scientists are generally sensitive to the temporal dependency of the phenomenon in which they specialize. Problems arise when different types of environmental scientist converse without clarifying the temporal scale in which each of them operates. Scale will be addressed separately below.

While most environmental scientists probably consider themselves firmly imbedded in the materialistic, Newtonian–Cartesian view of space and time, professionally engendered distortions and biases probably alter those views, at least in the experiential mode of cognition. Assuming that is true, we must be prepared to clarify the context of a particular problem so that all involved are attuned to the same phenomenon with the same senses of resolution, extent, and time dynamics. Scaling is part of the problem in good communication between environmental scientists, but not the entire problem. Deeper ontological mind-sets are involved.

3.5   Environmental ontologies in ecology

The deep-seated differences between environmental scientists of different disciplines are interesting and important, particularly because the propagation models presented in Part II are borrowed, to a large extent, from non-ecological environmental scientists. However, the remainder of this discussion will be focused on ecologists: a set of biological environmental scientists who differ among one another in ways related to the types of environment in which they work, and those places they know best. The focus will be narrowed further to those ecologists working primarily in terrestrial environments.

We need not scratch very deeply to find ontologic chasms between ecologists. Perhaps these differences seem greater to those, like the authors, who are embedded in this discipline. In any case, it seems that, in spite of apparently shared semantics, ecologists have only limited commonality in interpretations of ecological constructs such as "community," "ecosystem," "scale," or "level of organization" (McIntosh, 1985; Allen and Hoekstra, 1990). Contrasting definitions of ecology itself are symptomatic of these divisions. Haeckel originally defined ecology as "the scientific study of the interactions between organisms and their environment" (Begon *et al.*, 1986). Krebs defined ecology as "the scientific study of the interactions that control the distribution and abundance of organisms (Begon *et al.*, 1986). At one point, Eugene Odum defined ecology as "the science of the living environment" (Odum, 1959). The differences between these definitions are not trivial. They have large impacts not only on how ecologists communicate with one another, set scientific priorities, select questions for research, design sampling procedures or experiments, and interpret results, but also on the way that they conceptualize nature. These differences have many manifestations, one of which is how variable properties of nature are arrayed in space and time – a manifestation central to the subject of this book.

Ecology, as it has been practiced in the USA, has historically been divalent about whether nature is continuous or discrete in its organization. Both themes can be found since before the opening of the twentieth century, sometimes with the same authors at different times. On the one hand, a discretized view of spatial differentiation was inherited from early ecologists, who were often, but inconsistently, imbued with the localized, static community concept (Cowles, 1899; Forbes, 1925; Elton, 1927; Weaver and Clements, 1929; Clements and Shelford, 1939). On the other hand, views of continuous variation in space and time were being put forward in parallel. Cowles' (1901) monograph on how vegetation and associated geomorphological processes changed both continuously and discontinuously in the Chicago region over space and through time was a particularly lucid exposition of that viewpoint. Shortly after, Cowles' student, W. S. Cooper, created the lasting metaphor of vegetation being like an anastomosing river, ever-changing with time and conditions (Cooper, 1913). Other proponents of a continuous view of nature in its essential properties were Leopold (1941), Gleason (1926), Whittaker (1953), and Elton (1966) who worked on land, and Redfield (1958) and Margalef (1963) who worked on the seas.

Wiens (2000) described the waxing and waning of these perspectives in terms of recognition of environmental heterogeneity in ecological models. The intellectual environment after the Second World War reinforced an already present view of discretized nature consisting of homogeneous and temporally equilibrial communities or ecosystems. This ontologic view fitted the needs of theoreticians, whose point models and equilibrial assumptions came to shape the way that research sites were selected and then sampled, and how data were interpreted. By this philosophy, concrete examples of communities or ecosystems were to be delimited at those margins where the least interaction seemed to occur across space. In practice, sites or stands were mostly delimited on the basis of changes in vegetative structure. This view of nature as being composed of contiguous, discrete and non-interacting units was particularly convenient in the agriculturally gridded midwestern USA, where, in fact, much early thinking in ecology took place. Remnant forest stands, individual fields, tightly demarcated riparian zones, and well-bounded lakes and wetlands contributed to this ontology of a discretized nature. Obviously, discretized real estate with ostensibly minimal interactions between the discrete units produced an uncongenial environment in which to promote concepts such as propagation of influences across space. Such propagations were to be minimized, not appreciated.

Wiens (2000) credited theoreticians such as MacArthur (1967), MacArthur and Levins (1964), Levins (1968, 1970), Horn and MacArthur (1972), Levin and Paine (1974), and MacArthur and Wilson (1967) for creating the paradigm shift in the 1960s and 1970s that promoted incorporation of spatial and temporal heterogeneity into ways of conceptualizing nature and carrying out ecological research. It is ironic that after decades of focusing on discrete and independent community and ecosystem units, exchanges between patches became an intensely active subject of ecological research. In this sense, one ecological ontology metamorphosed into another.

The ontologic shift described above was important for those who experienced it, but, in fact, the theoreticians listed by Wiens primarily addressed animal populations and communities. The underlying divisions of ecology were such that this paradigm shift had limited influence on those ecologists who studied the environmental physiology of plants and animals, or studied vegetation. Among plant ecologists addressing the venerable problems of the nature of vegetation, an independent and parallel controversy brought about a similar turn of events. The formalisms of early plant community ecology, and thus understandings of vegetation, became dominated by the concept of discrete spatial units having identity, in site quality or successional terms, with a standard for a region: the climatic climax (Weaver and Clements, 1929). By implication, constituent species had either a high degree of mutual facilitation or had similar ecological requirements.

This spatially discrete and compositionally typal philosophy was attacked to little immediate effect by Gleason as early as 1926, but with more impact by Whittaker in the 1950s (cf. Whittaker, 1953) and finally by the "continuum" school of plant ecology originating primarily at the University of Wisconsin in the 1960s (Curtis, 1959). This last viewpoint was based on perceived differences in niche requirements of species, in competitive rather than facilitative interactions between species, in the strong role of disturbance and stochastic dispersal events in shaping the composition of vegetation at particular sites, and in the continuous changes in species composition displayed along environmental gradients unless environmental discontinuities were encountered. The conflict between the typal and continuous nature of vegetation reached its climax when in an unusually critical paper, Daubenmire (1966) articulated the typal view, to which Vogl (1966) and Cottam and McIntosh (1966) replied with equal fervor. To an extent, this debate was a case of inadequate specification of scale in promoting basic ideas. Nevertheless, it is a classic case of clashing ontologies about nature. This clash did not provoke a broadly accepted common ontologic mind-set among plant ecologists. The two viewpoints are articulated and used by different practitioners and under different circumstances. For some plant ecologists, dual ontologies about the nature of vegetation live side by side in their minds and are used flexibly for different purposes.

3.6   Ontologies of landscape ecology

Nowhere in ecology has the representation of environmental space been addressed more vigorously and rigorously than by the ecological subdiscipline landscape ecology. Landscape ecology has several intellectual traditions of its own (Moss, 1999; Wiens, 1999) and we are representing only one North American tradition, most fully manifested in Turner *et al.* (2001) in this discussion.

According to this tradition, landscape ecology "deals explicitly with the causes and consequences of spatial patterns in the environment, [and] with the effects of spatial pattern on ecological processes" (Wiens, 2001c). Landscape ecology is primarily a terrestrial pursuit, as its name indicates. As such, it focuses on patterns on land surfaces, although those patterns may include streams (Malanson, 1995; Ward, 1998; Wear *et al.*, 1998). Otherwise, application to lakes and oceans has been limited mainly to two-dimensional surface features rather than volumetric environments (Wright and Bartlett, 1999). Three-dimensional analyses are carried out; however, interpolation between data points is difficult over the vast volumes involved in oceans (Greene *et al.*, 1994, 1998; Benfield *et al.*, 1998).

Writing in the context of terrestrial landscape ecology, Forman and Moore (1992) recognized that there can be more than one way to represent spatial heterogeneity on land. They asked, "Does one focus on (1) a boundary-less pattern of gradients (analogous to certain impressionist paintings), (2) patches in a mosaic, (3) a network of corridors, or (4) boundaries or edges in a mosaic?" They admitted that all approaches lead to understanding but came to the conclusion that the last option "leads to deeper insight in the structure and functional role of boundaries." Of course, insights on boundaries can only be made where boundaries exist, particularly at the scale and resolution in which the work is being performed (Turner *et al.*, 2001). By deciding that boundaries exist, one defines an environmental ontology that is critical to how nature is visualized and field research is performed. The ways by which spatial heterogeneity is conceptualized and represented is central to the ontological vision of landscape ecologists.

Wiens (1992, 1995, 2000) has addressed the spatial conceptualization of land surfaces by landscape ecologists in terms of "ecological heterogeneity." Wiens recognized four kinds of spatial heterogeneity, which he said formed a graded series from spatially implicit to spatially explicit or locational (Fig. 3.5). The first he termed "spatial variance," meaning the statistical measure of aggregation among objects in a two-dimensional field. His second type was "patterned variance," in which objects in the field are statistically characterized by similarities among objects near to one another and dissimilarities between objects distant from one another. Neither of these kinds of heterogeneity is locationally explicit, but merely statistical characterizations of variation in space. Wiens' third type was "compositional variance." This is a spatially explicit representation of objects in a field that vary qualitatively from one another. The final type was termed "locational variance," in which objects having definite shape and area (polygons or contiguously clustered raster cells) are explicitly located. Such polygons are typically referred to as "patches." The property represented by the patch can be many things but, in practice, is often a physiognomically

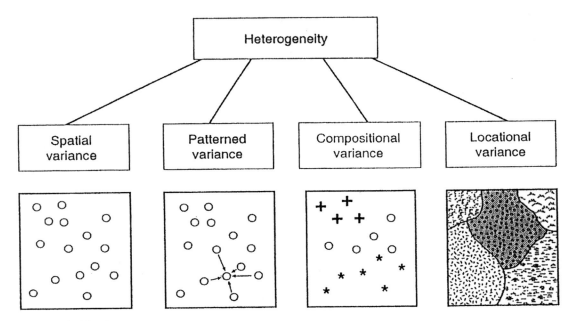

FIGURE 3.5
Kinds of ecological heterogeneity according to Wiens (2000). (Reproduced with
permission from the author and Blackwell Science.)

defined unit of dominant vegetation. Not included in Wiens' system is a quasi-
continuous system represented by grids (rasters) in which grid cells might be
members of classes or have infinitely variable scores for properties of interest
(Fig. 3.5).

Patch representation of environmental heterogeneity is the one most commonly used in landscape ecology and there are several elaborations of this
approach. The first is the patch matrix, in which patches are imbedded in a
homogeneous matrix dissimilar from the patches (Fig. 3.6A). The second is
the patch mosaic, in which the area is divided into contiguous patches with no
intervening matrix (Fig. 3.6B). The third is more of a switch in emphasis than
a change in structure. It is the patch matrix or patch mosaic with "ecotones"
(boundary zones between dissimilar patches) or "corridors" connecting, usually elongated, patches (Fig. 3.6C). Yet another spatial heterogeneity format
is the raster grid (Fig. 3.6D), in which grid cells can be treated as members of
binary units, a limited number of classes, or with infinitely variable values for
the property of interest. Grain is especially critical in giving the grid system
a degree of continuity. If the grain is too coarse, a grid is just a special kind
of patch system. Resolution, or grain, is critical to the nature of representation of heterogeneity in nature (Turner *et al.*, 2001). Usage of a patch-matrix or
patch-mosaic conceptualization of nature dominates on land and the marine

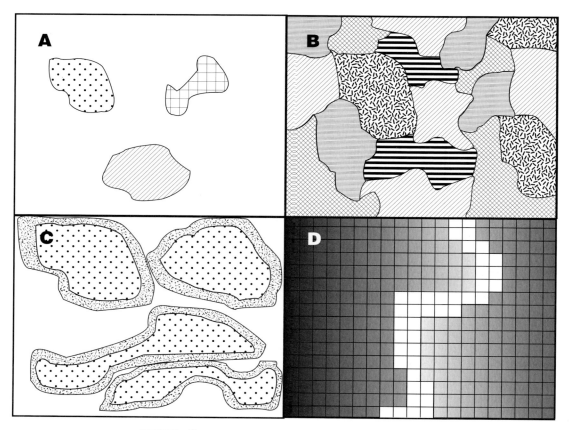

FIGURE 3.6
Four methods of patch representation in landscape ecology. See text for explanation.

littoral zone but not in the pelagic zones of lakes and seas, where continuous fields make more sense.

Patches can be defined by any property of the environment, and their scale and resolution are set as is appropriate for the phenomenon in question. In practice, patches are typically defined by land management units or units of homogeneous vegetation. Therefore, they most often are delimited at a scale of human perception based on features that are seemingly stable within the human time perspective. As such, patches are, without doubt, convenient and intuitively agreeable representations for permitting humans mentally to organize the spatial heterogeneity of nature. Patches are also more prominent in areas undergoing recurrent, exogenous disturbances (Noble, 1999). Certainly, a patch-mosaic viewpoint matches our human perception of some landscapes, particularly landscapes in human-dominated terrains as found in western Europe or the midwestern USA.

There are other places, though, where it is more difficult to envision a patch mosaic, and defining patches where they are indefinite at the scale of choice is an arbitrary process. Illustrating the difficulty of these environments for those committed to discrete representations, Haines-Young (1999) described these as "fuzzy landscapes." Examples might be unbroken tropical or temperate deciduous forests, chaparral (within historical fire boundaries), or alpine tundra. In environments where variation is more continuous, it is more recognizably arbitrary to define a patch mosaic without making a large switch in scale. Consequently, one particular conceptualization of nature works better in some places than in others.

Aber *et al.* (1999) recognized that there is more than one way to view landscape heterogeneity but felt that most landscapes could be described as patch mosaics and that such a structure was preferable for a number of practical reasons. Reynolds and Wu (1999) said that if patches were not obvious, then the environmental data field needed to be examined at another level of resolution – what they referred to as scale – to identify them. Later in their paper, Reynolds and Wu finessed this issue by saying, "While one may argue about whether these units (patches) are inherent entities or results of interactions between human perception and the objective world, in either case explicitly identifying these hierarchical levels is essential in order for us to simplify and understand functioning in complex landscapes." Clearly, these two authors were wed to patches whether they were appropriate or not, a symptom of fixed ontologic views of nature. This patch description and definition of landscapes tends to be incorporated into ecosystem-level interpretations of landscapes by Chapin *et al.* (2002).

A problem with any representation of spatial heterogeneity is whether the criterion used to demarcate the heterogeneity is actually related to the phenomenon in question. It is easy to create a patch mosaic out of a forest composed of regrowing logged blocks, or out of a system of agricultural fields. These kinds of heterogeneity are easy for humans to see and to fit our own operational scale for human endeavors. However, these units may not be germane to some processes, either because they have limited functional relationship with the mechanics underlying the process or because they are at the wrong scale. For example, a forest patchwork might be mechanistically appropriate for defining the movement of small mammals but would have no, or only marginal, value for describing hydrological flows. The criterion for defining environmental heterogeneity must not devolve from what is most apparent to the human eye but must arise from what is most instrumental in controlling the phenomenon being addressed (Allen and Hoekstra, 1990). Related to this issue is the fact that environmental phenomena do not necessarily map on the same spatial features, nor do all spatial features map on top of one another. In fact, different spatial

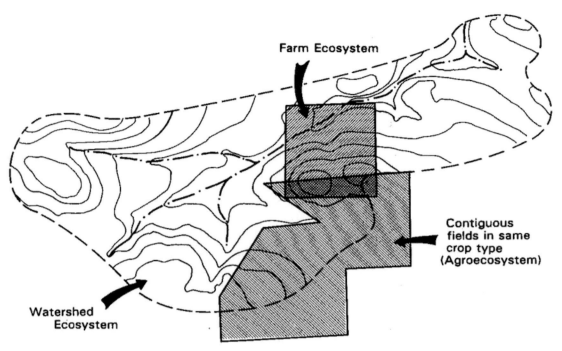

FIGURE 3.7
Ecosystems and landscape patches of the same scale are bounded differently,
depending on essential criteria set by the phenomenon in question. Farms, as
economic units, are bounded differently from contiguous fields, having the same
management practice, and from hydrological units, having a common drainage.
(From Reiners (1995). Reproduced with kind permission of Kluwer Academic
Publishers.)

features may be quite incongruent in terms of appropriate resolution or boundary location (Fig. 3.7). Vegetation patterns, such as fields or woodlots, are likely
to map independently from watershed boundaries, soil types, or availability of
critical habitat elements like standing dead trees.

Representations of single patterns of heterogeneity to describe or analyze
processes imply that the properties of those pattern units are the only ones controlling spatial variability in that process. In many cases, two or more measures
of spatial heterogeneity are needed to understand those processes. Separate spatial representation of food source and of concealment or reproductive sites may
be needed to understand the distribution of some animals. For example, sage
grouse habitat is defined in multiple terms. Vegetation cover requirements in
winter are different from those in the summer; distances to lek sites and reliable
summer water are also critical. More than one spatial representation is necessary to understand the distribution and abundance of this species, a situation
likely to be true for many, if not most, ecological phenomena.

Time must also be considered, along with spatial heterogeneity, for representing processes in the environment. Davis (1986) once defined communities as: "A local community – to a first approximation, and viewed over a sufficiently long time span – is an ephemeral ensemble of species that originated somewhere else." Davis was considering changes over time frames of centuries to millennia but this ephemerality, or possibly seasonality, applies to other phenomena. For example, exchanges of water and energy with the atmosphere will be influenced by leaf-on/leaf-off conditions – conditions that change annually. In such a case, the controlling, and spatially distributed, factor may shift from vegetation patches to soil types, requiring a shift in spatial representations at appropriate times of the annual cycle. Snow-on/snow-off is likewise important to the behavior of some animals. Wiens (2001c) illustrated the importance of change during the successional time frame with respect to animal dispersal. Clearly, the time dimension complicates our conceptualization of nature.

By deploying any discrete representation of spatial heterogeneity – like patches – to represent a process, one makes a philosophical statement that nature can be correctly, or at least adequately, envisioned as, and defined by, polygonal land units that are homogeneous within and distinct from the surrounding matrix or contiguous patches. Such a practice is a personal or, perhaps, disciplinary ontologic statement about nature. One must ask whether devotees of patch-based representations only work in discretized landscapes, or are they only able to see nature in those terms? A patch-based representation of environmental space has operational advantages only if:

- the criterion by which the patch is defined is critical to, or controlling, the phenomenon of interest,
- the patches used are delimited at an appropriate scale over an adequate extent of environmental space (Wiens, 2001b), and
- the patches themselves are not in flux during the time frame for which a time-dynamic phenomenon is being analyzed or simulated.

If these advantages are not met, then a patch-based system may not be the best representation of environmental space, and persistence in its use may be symptomatic of entrapment into a particular ontologic view of nature.

3.7   The central place of scale awareness in environmental science

As stated above, recognizing the appropriate scale, or scales, for representing structure or process is crucially important. Much has been written on this subject but, like virtue in public affairs, there can never be too much scale awareness in environmental science and ecology. In fact, O'Neill and King (1998) have published a paper (in a book on scale) titled, *Homage to St Michael: or, why are there so many books on scale?* In spite of this outpouring, confusion about scale persists.

Wiens (2001b) provides an excellent review of the many meanings of scale and its importance in experimental ecology.

> I propose that, in an ecological context, "scale" should refer to a quantified portion of the spectra of space or time. This portion . . . is bounded by *grain*, the resolution of measurements, and *extent*, the upper limit of the range of scales over which measurements are taken. "Scaling," then, involves relating measurements made at one scale to those made or predicted at another (i.e. shifting grain and extent to encompass a different portion of the time or space spectrum). The scales at which measurements are taken in space or time are absolute metrics . . . although references to "fine" or "broad" or "short" or "long" scales are as relative as "small" or "large".

Wiens' definition will be the one adopted throughout this book. It is clearly the opposite meaning to that used by cartographers, and it most definitely is not a level of organization, a logical error energetically exposed by Allen and Hoekstra (1992), O'Neill and King (1998), and Allen (1998).

Given the avalanche of literature on this topic, only two points need to be made here. First, measuring, analyzing, and representing environmental processes at the appropriate scale is essential for the success of a scientific endeavor (cf. Ludwig *et al.*, 2000). So many mistakes and misunderstandings in ecology have originated from misunderstandings of spatial and temporal scale that perhaps ecologists should, as a practice, define their scale when engaged in conversation or writing, just as they set font type and size when engaged in word processing. It should not be assumed that there is a default scale, like a "human scale," that applies to every phenomenon. Spatial–temporal phenomena can be analyzed on a single leaf where resolution may be less than a millimeter and extent a few centimeters. It can be analyzed on a ground plot where grain is centimeters and extent meters. It can be analyzed on units of terrain with extents of kilometers or thousands of kilometers, or it can be analyzed over the entire globe, a surface area extent of 510×10^6 km^2.

The second point to be made about scale here is that assumption of a default scale in ecology results from the fact that ecologists tend to see nature as humans, not always as scientists. This does not necessarily lead to good science. The scale of familiar places and processes can be as great a trap to good science and effective communication as are the all too familiar ways of depicting spatial heterogeneity. Temporal and spatial scale should be joined together with kind of imagery, choice of surface versus volumetric representation, assumption of natural boundaries, representation of spatial heterogeneity, and choices of controlling criteria as key factors shaping cognitive understandings of nature by sociocultural scientific "clans."

3.8   A hydrodynamical concept of environmental space

There is no "best way" of conceptualizing space for all questions and all situations. However, as is true for many scientific questions, it would be best if we approached any spatially distributed phenomenon with a minimal set of preconceptions about how nature is organized in space. Our thoughts here are an extension of Gardner and O'Neill's (1991) argument for a "neutral model" for landscapes; a useful but relatively narrowly conceived concept generally applied to stochastic, constructed landscape models (Keitt, 2000) rather than to actual environmental domains in which real features must be represented. The suggestions that follow apply in two ways. The first way is in striving to separate our own personal conceptualizations of nature from the best conceptualization for the phenomenon in question. The second way is to help to abstract and represent the best conceptualization for spatial analysis or for spatially distributed process modeling.

We suggest starting out with an open mind towards the proper scale of the environmental domain and the objects of interest within it. Imagine the defined environmental domain, a section of "nature," to be of any possible scale, from a leaf to the globe. Start with a definition of scale that matches extent and resolution or grain (Turner *et al.*, 2001).

A second consideration is number of spatial dimensions. For some purposes, nature can be adequately represented in zero dimensions (a point), for others one dimension (a gradient), for others two dimensions (a surface). However, for some domains, a third, depth dimension is imperative for representing features of heterogeneity. Two examples are representing the light environment throughout canopy systems at all scales (Moffett, 2000) and describing particulate organic matter changes with depth in a lake or ocean. These require either that thematic layers have a depth representation incorporated into each grid cell for the two-dimensional area or that for each coordinate position for each grid unit in the domain, there are multiple layers of defined depth representing the same attribute: in effect, three-dimensional pixels or "voxels" of continuous vector space. In fact, all ecological phenomena, even the movement of slime molds, operate in three dimensions, but we can squash representation of some phenomena down into two dimensions with lesser loss of realism than with others.

A third consideration is whether representation of structural heterogeneity in a domain should be discrete or continuous. This, of course, is itself a scaling issue. With finer and finer resolution, features that were once discrete become continuous. At a particular scale, one grain size might make some features discrete while others are continuous. For example, a plant leaf domain resolved at millimeters might have stomatal distributions represented discretely but vestiture density as continuous. As another example at the kilometer scale, road networks and dwellings might be discretely represented while topography is continuous. Of course, any discrete features can be represented continuously by using grid cells small enough to subdivide the otherwise discrete feature.

Whether they are leaf hairs or roads, features defining heterogeneity in an environmental domain belong to a common class, each of which may be considered as a spatial field of potentially independent variables. In GIS parlance, these fields are referred to as "themes" or layers, each of which can be represented as a separate spatial rendering in digital form for the domain. The domain can be thus decomposed into as many thematic grids as are of interest for analysis (cf. Aber et al., 1999). Solon (1999) considered each thematic grid as being equivalent to his "partial geocomplex."

A fourth consideration is how qualities of each grid cell or discrete feature vary in time. The rate of temporal variation can, itself, vary on its own time scale (Allen and Hoekstra, 1992). Human structures can come and go over a century but topographic change might only be perceptible over millennia. Consequently, if class values were represented as colors on the fine-grained grid, the colors in a single thematic layer might be twinkling from one to another at different time scales. To the extent that there are transfers of materials, energy, or organisms between landscape features, discrete or continuous as they may be, such transfers can also be represented as occurring between the grid cells according to the rules directing such movements.

Based on these ideas, an extreme example of an abstract conceptualization of a defined domain would be composed of multiple layers of several-to-many defined features in three dimensions in a continuous, very fine grid system. It would also include the potentiality for change in class attributes for every grid (or cube) unit in this complex, and with the potential that every grid unit in every layer could influence another grid unit by some kind of defined rules. This ontologic framework for conceptualizing nature starts with an assumption of continuous change but the framework may be partly or wholly discretized if features are, indeed, discrete at the appropriate scale, and if discretization contributes toward understanding the phenomenon being addressed. The second leading feature of this construct is one of temporal fluidity in structural change and movement. The emphasis is on continuity and change rather than partitioning and stasis. In fact, this is the view oceanographers must have of the sea, a continuous fluid of vast three-dimensional character (Greene et al., 1994, 1998; Levinton, 1995; Benfield et al., 1998; Cowen et al., 2000).

   We recognize that, at some point, such a hydrodynamical mental construct of nature in time and space is impractical. Practically speaking, we must have maps to do our work, however artificially they may represent nature and as limited as they are to particular time frames. Operationally, to simulate transport processes across domains, we have to simplify and, usually, to discretize in order to represent nature for computational purposes. Often, patch representations are not only convenient but also conceptually appropriate and, certainly, computationally more efficient. The hydrodynamical mental construct may be an ideal, neutral perspective for gaining initial understanding of structure and function in spatial domains, but finally, the system will have to be simplified as needed. In fact, readers will find that representations of propagation processes demonstrated in the second part of this book are very much simplified compared with this idealized hydrodynamical model of nature.

Representing and predicting propagation phenomena: modeling in explicit–realistic space

The mere formulation of a problem is far more often essential than its solution, which may be merely a matter of mathematical or experimental skill. To raise new questions, new possibilities, to regard old problems from a new angle requires creative imagination and marks real advances in science.

Albert Einstein

4.1 Definition and description of geographic information systems

The models described in Part II use geographic information systems (GIS) as a means for organizing data and displaying model outcomes in spatial representations. While GIS makes this work possible, it also places constraints on the way we can visualize nature. It is, therefore, important to know something about how GIS works and its limitations.

Clarke (1997) has pointed out that GIS is at once a toolbox, an information system, an approach to science, and a big business. In all of these forms, it is built on spatial data. In its toolbox manifestation, Clarke described GIS as an "automated systems for the capture, storage, retrieval, analysis, and display of spatial data." As an information system, Clarke followed Duecker (1979) who said,

> A geographic information system is a special case of information systems where the database consists of observations on spatially distributed features, activities or events, which are definable in space as point, lines, or areas. A geographic information system manipulates data about these points, lines and areas to retrieve data for ad hoc queries and analyses.

As an approach to science, Clarke notes the spectacular mutual reinforcement of several spatial technologies on one another. These technologies include GIS, remote sensing, the global positioning system (GPS), computing, and communication. They have been combined to become what some have termed "geographic information science," sometimes symbolized as GISc. Goodchild (1992) defined GISc as "the generic issues that surround the use of GIS technology, impede its successful implementation, or emerge from an understanding of its potential capabilities." Couclelis (1999) noted that GISc is a "meta" science in that it is not really about the geographical world but about information on the geographical world.

Finally, Clarke reviewed how GIS has become a multibillion dollar business: over the first half of the 1990s, GIS saw a double-digit annual growth leading to a total value of about US $6 billion a year by 1997. This is an aspect of GIS that is peripheral to this book but central to career aspirations of many students. For other definitions of GIS, see Maguire and Dangermond (1991).

GIS is a field in itself with its own courses, journals, and societies (Clarke, 1997). It has multiple personalities in line with its multiple definitions and uses. It is an area of intense and fascinating intellectual inquiry in and of itself. At the same time, it is often, and erroneously, viewed as merely a computer-based technology that can be "learned" by mastery of a piece of software. In the context of business, training, and employment, it is viewed as a burgeoning industry. In reality, GIS is a combination of all these viewpoints.

For the most part, GIS has inherited and promotes the treatment of space in Newtonian terms using Euclidean analytical and computational geometry. It rests on the conviction that the world can be perceived by observations: the materialistic or empirical–analytic view of reality (Raper, 1999). Inescapably, it has inherited the cartographic tradition of a map view of the world. This has given GIS cartographic conveniences along with some weaknesses, including the fact that "maps are static, flat, 2-dimensional, precise, and not well suited for conveying the fact that the level of knowledge or certainty over their range is often far from uniform" (Couclelis, 1999). Couclelis noted, for example, that relationships like flows and interactions are almost always n-dimensional and the usual two-dimensional, map-like representations provided by most GIS are inadequate for presenting relationships between entities on the modeling domain. The topology inherent in GIS dominates over the nature of the relationships we attempt to represent. For example, most ecological relationships involve three-dimensional space but are flattened into two dimensions in a GIS. This is partly the result of acceptance of two dimensions as adequately representative of phenomena of interest. It is also because of lack of familiarity with three-dimensional GIS software, and with the complexity of operating in three dimensions. Three-dimensional software packages are available, usually as raster representations constructed from "voxels" in the place of "pixels" (Raper, 1989; Raper and Kelk, 1991; Raper, 1999). It is important that users do not allow the flattened, two-dimensional projections of ecological processes to become another limiting "reality" in the mind of the researcher. It is as easy for users to slip into seeing the world as irregular triangles, grid cells or hexagons as it is to see everything as patches.

4.2   GIS data modeling

Described in a very general way, GIS software records the locations of points, lines, and areas as collections of x,y coordinates. However, it is the ability to manipulate these entities and compute relationships among them that gives GIS its power. To a major extent, the property making these functions possible is inclusion of topology in the relationships between these locational entities. It is topology that makes GIS more than a mapping tool, thereby setting it apart from computer-assisted design (CAD) and computer-assisted cartography. Clarke (1997) defined topology as:

> (1) The property that describes adjacency and connectivity of features. A topological data structure encodes topology with the geocoded features. (2) The numerical description of the relationships between geographic features, as encoded by adjacency, linkage, inclusion, or proximity. Thus a point can be inside a region, a line can connect to others, and a region can have neighbors. The numbers describing topology can be stored as attributes in the GIS and used for validation and other stages of description and analysis.

Topology allows for error detection of GIS data by providing a check on consistency in locational attributes relative to other features. This is an important asset for those building data or using it. Further, topology built into an initial spatial data set provides a template for overlying other digital maps of the same area so that differences in delineations can be easily detected and reconciled. This means that analyses of digital map features can be conducted without reference back to the x,y coordinates used to build the original digital data. It is important to remember that GIS data include topological relationships as well as the more evident attribution data for data features. Topological relationships allow the GIS to perform analyses such as determining shortest flow paths, which cannot be done, or at least done as efficiently, with a non-topological database.

The way data are structured, visualized, and analyzed is based on how we conceptualize the external world and, therefore, derives from the user's conceptualization of nature. However, GIS technology itself may limit how data are structured. Understanding GIS data models is one way to approach this problem. According to Raper (1999), "... a data model [is] a set of guidelines for representation of the logical organisation of the data in a database consisting of named logical units of data and the relationships between them." A typical GIS data set spatially represents direct observations of position, form, or

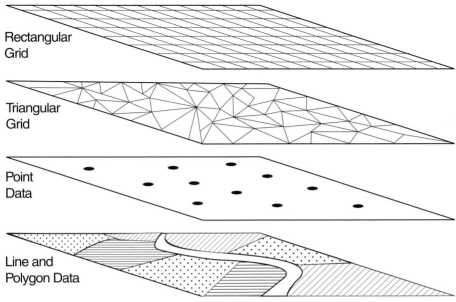

FIGURE 4.1
Hypothetical graphical representations of digital data in four thematic data sets, or layers from a geographic information system. With topology, the connectivity between features within data sets is incorporated along with the locations of the features. Topology also provides coincident location between locations of the different thematic layers.

behavior for geographic objects. The objects could be zero-dimensional points, one-dimensional lines, two-dimensional areas, or three-dimensional volumes (Raper, 1999). Only data describing a specified class or GIS "theme," such as locations of streams, roads, or nesting sites, are entered into a single data set. A GIS stores "thematic" data separately but also allows comparison and analysis using multiple themes simultaneously (Fig. 4.1).

Very generally, there are two approaches to organizing spatial data: vector and tessellation. **Vector** data structures represent entities as *x,y* coordinates of points, lines, areas, and volumes to make representations (Fig. 4.1). **Tessellation** data structures, usually used with continuous data, describe geographic features in terms of polygonal units in a matrix that can be called a mesh, lattice, or array. **Regular** tessellations are usually based on rectangular grids and are composed of what we normally think of as pixels (Fig. 4.1), whereas **irregular** tessellations can be composed of irregular polygons such as regular or irregular triangles, net elements, or hexagons (Fig. 4.1). Quite frequently, regular tessellations are structured as equal-sized squares, or pixels, and are referred to as a raster data structure (Maguire and Dangermond, 1991; Burrough and McDonnell, 1998). Data modeling in regular raster arrays is so common that frequently the term raster is used rather than tessellation in discussions contrasting tessellation with vector data structuring. For further reading on GIS data structuring, see Clarke (1997); for particular consideration of three-dimensional environments as met in marine environments, see Li (2000).

Object-oriented GIS languages now make it possible to model objects without the constraints of vector or tessellated data structuring; however, the terms "object-based" or "object-centered" are nebulous through their varied usages (Burrough and McDonnell, 1998). The goal of object-based modeling is to describe an object in a database in terms of its attributes (state) and of procedures describing its behavior (operations). Object-centered software permits the user to define an object, such as an individual animal, and give it behavior for movement in two-dimensional or three-dimensional space. Even so, for the object to interact with its environment, it must be related to an external representation of that environment using conventional GIS.

GIS is primarily concerned with organization, visualization, and analysis of spatial relations. Only secondarily is it designed to address the temporal nature of events. Raper (1999) stated,

> Time is usually assumed to be absolute, operating as a frame of
> reference where events partition a single universal timeline. By
> convention, representations of space in two or three dimensions are
> realised at an instant in time creating "timeslices/time volumes" which
> can be operationalised as layers or volumes. Changes can then be
> defined as geometric differences between "timeslices/time volumes."

By arranging timeslices in rapid sequence through animation techniques, dynamic behavior can be visualized. Dealing with time is a research problem for the GIS field. Ideally, GIS could be operated in four dimensions with time as the fourth dimension (Rogowski and Goyne, 2002). Peuquet (1999) has reviewed some novel conceptualizations for dealing with temporal data in a spatial context, but widely accepted and generally applicable approaches have yet to be achieved with commercially available software (Couclelis, 1999).

4.3   Process modeling in GIS

GIS can be used for visualization (making maps and digital imagery), for analysis of spatial relations (determining conditions, trends, routes, patterns), for making new syntheses of spatial relations (quantifying and displaying relationships resulting from analyses), and as a platform for process modeling. GIS is not required for spatially distributed process modeling. In fact, some would say that it has more disadvantages than advantages (Fischer, 2000; Fotheringham, 2000). The justification for using GIS rests, in part, with the degree of integration of the model with the GIS, the complexity of the model computations, and whether the GIS has some built-in functions that are useful to the modeling process (Spiekermann and Wegener, 2000). For those not deeply committed to spatial modeling but familiar with GIS operations, a GIS platform is at least a convenience. Experts believe that GIS packages will become increasingly attractive for spatially distributed process modeling as they develop to incorporate more modeling functions. The example models described in Part II of this book are linked to a popular GIS platform. The GIS is used here (as with many other cases) to provide the underlying thematic databases and for providing a spatial visualization of the model output at the conclusion of the model run.

Process modeling is the representation of the dynamic behavior of a continuous, "real" physical or biological system over space and/or time. In contrast with empirical models, process models are based on understanding of physical, chemical, geological, and biological processes through their mathematical descriptions (Mitasova and Mitas, 1998). In fact, many process models are mixtures of mechanistic representations with empirical relationships. Originally, process models were developed for projection over time, but software implementations have made possible projection over representations of space. Such models are referred to as spatially distributed process models.

Mitasova and Mitas (1998) have organized properties of spatially distributed process models in a way particularly suitable for this book, and their organization and definitions will be followed here (see also Wegener (2000) for an alternative approach). According to these authors, process models can be classified as listed in Table 4.1.

Process models can also be classified according to the modeling approach upon which they are structured. Mitasova and Mitas (1998) divide process models into deterministic, stochastic, rule-based and multiagent sets. Deterministic models feature unique inputs leading to unique solutions for well-defined linear models, and with multiple outputs possible for non-linear models. Stochastic models simulate spatial–temporal behavior of phenomena with random components (Mitasova and Mitas, 1998). This means that, while unique inputs lead to unique outputs for each model run, sufficient multiple runs can generate probability distributions of outputs. Random inputs can be supplied from one or more than one input variable, so that a joint frequency distribution incorporating measures of covariance can be generated (Reiners *et al.*, 2002).

The spatial distribution of either deterministic or stochastic model equations is solved using various numerical methods. Often, these are finite process models that operate recursively across tessellated spatial arrays, so time-varying behavior is produced. Those models based on raster grids are termed "finite difference" models; those based on other tessellations such as triangulated

TABLE 4.1. *A taxonomy of spatially distributed process models*

Classification group	Examples
Area of application	Natural processes
	Socioeconomic/anthropogenic processes
Type of spatial distribution	Homogeneous, spatially averaged units such as watersheds, counties (lumped models)
	Fields/multivariate functions discretized as tessellation units
	Networks (systems of nodes and links)
	Points representing individuals and agents
	Combinations of fields, networks and points
Nature of spatial interaction	None: location dependent behavior only
	Short range: close neighborhood interaction
	Long range: expanding interaction
Type of process	Fluxes
	Proliferation and decay
	Population dynamics
	Intelligent agents: systems of independent entities that interact between themselves and the environment with a certain degree of decision-making capability
Spatial extent	Local
	Regional
	Global
	Multiscale or nested models

From Mitasova and Mitas (1998).

irregular networks or as "tuples" or "cohorts" of cells having identical properties in all thematic data files are referred to as "finite element" models. Both of these "finite" forms of spatial representation specify a limited number of topologically closed areas for which states of a physical system can be computed and stored. These finite process models are limited by the number of interactions permitted across the boundaries between grid cells or triangles and by the fixed scale of the spatial representation (Raper, 1999).

An alternative approach to model operation in a tessellated field is termed "path simulation," also known as "path sampling," "random walks" (Mitasova and Mitas, 2002), or "transition pathways" (Chandler, 1998) in which movements can be calculated with or without dependency on a mesh. In this approach, originated for physics and chemistry problems (Gardiner, 1985), objects or agents move independently of any tessellation by following rules describing their movement in an appropriate field space. These rules can take the form of operators, which usually have a probabilistic component. By simulating movement iteratively in a Monte Carlo fashion, results can be combined and displayed as a probability field. Chandler (1998) used the metaphor "throwing ropes over rough mountain passes in the dark" in his application of this process for defining energy or reaction fields for chemical processes. This approximative approach yields a continuous field of conditions or flux rates that is highly expedient for modeling surface fluxes. Mitas and colleagues (Mitas *et al.*, 1996; Mitas and Mitasova, 1998a,b) demonstrated this approach for modeling erosion and deposition of sediment.

Many spatially distributed models describe dynamics across the entire domain field. Alternatively, flows can be confined to prescribed or predicted flowpaths such as road systems or stream drainages. Collectively, these are referred to as network models. Networks are defined by the topological relationships between rigid, multisegment systems of the network and are a subset of finite element models. The connectivity between the segments allows for flow (movement) to be modeled. Network elements can have an associated impedance that determines the rate at which an entity can move through that element. Examples include speed limits on highway networks, bed friction in stream networks, or soil porosity in subsoil water flow networks. Networks can be created using any of the data structures (vector, tessellation, object oriented), although most often the vector model is used.

The third general class of model described by Mitasova and Mitas (1998) is term "rule-based." In these models, local objects are governed by rules as "agents" or "cellular automata." This class of model is also sometimes referred to as "agent-operated." Such models involve updating states or values of domain elements such as grid cells at discrete time steps according to transition rules applied universally and synchronously to each cell at each time step. Cell values are determined based on a geometric configuration of neighbor cells, which are specified as part of the transition rule. The application of simple rules applied to microscopic level data with cellular automata can generate simple, holistic structures and behavior (Openshaw, 1991). Microscopic level data would in this instance be the raster pixels of the finite element data structure, or other tessellations underlying the finite element data structures. Raper (1999) noted that the knowledge programmed into the agent "is implicitly and explicitly 'local' in nature since no agents have access to the global state of the system." In aggregate, however, complex global behavior may emerge from application of relatively simple rules to the cells or agents of the model.

The fourth class of model, according to the Mitasova and Mitas (1998) classification system, is the "agent-based simulation" approach (in contrast to the "agent-operated" class of models just described.) This approach is commonly referred to as "object-oriented," or as "individual-based" modeling by other authors. In these types of model, the agent, or multiple agents, is any actor in a system that can generate events that affect itself and other agents. Typically, an agent follows a set of rules describing its responses to stimuli and other agents. Inasmuch as the rules are imbedded in the agent itself, this is different from the agent-operated models described above, which are dependent on attributes of any particular raster or vector field. Objects can move freely in two- or three-dimensional space based on the operations built into the object's functionality.

Process-based models can describe phenomena at several levels of temporal variation including steady state, time series or dynamic (Mitasova and Mitas, 1998). Steady state has no temporal variation and is useful for diagnostic applications. Time-series outputs are computed by running steady-state models with time series of input parameters. Dynamic outputs incorporate some form of time-varying input variables, either extrinsic or intrinsically generated, and, generally, time-varying output variables. According to Wegener (2000), if models are organized with discrete time intervals, they are termed "simulation models." If they have fixed time intervals or periods, they are termed "recursive," and if they have variable time intervals, they are termed "event-driven."

One of the problems of simulating environmental processes is that multiple scales may have to be involved for effective modeling. Critical phenomena at some points in the environmental domain may require finer-scale modeling than for the domain as a whole. In such cases, a system of nested models might suffice. It is possible that recursive transfers from high-resolution models to coarser-resolution models may have to take place to set appropriate boundary conditions for processes that manifest themselves across a series of scales (Mitasova and Mitas, 2002). For example, a population explosion of a forest pest insect may initate locally and require fine spatial–temporal resolution to be properly modeled, but the dispersion of adults and consequent establishment over a broader domain may require a coarser resolution modeling system so that the procedure does not break down with excessive computational load or computing time.

Implementation of process modeling with GIS functionality can take on several forms. In some cases, the process model can be fully integrated into the GIS (imbedded coupling) and be operated with application programming interface, scripting language, or map algebra operations. In this case, data import is required and the model is run as a GIS function or command. A lesser degree of integration is achieved with a system Mitasova and Mitas (2002) termed "integration under a common interface," or "tight coupling." In this case, the process model is developed outside of the GIS but linked through a common interface that guides the user through sequential steps of data input, model operation, and subsequent analysis and visualization. Yet another system with looser coupling is possible by keeping the model and GIS independent of one another and simply importing data from the GIS to the model through individual commands and returning the flow of model output back to the free-standing GIS. Finally, the model can be developed independently and given its own GIS functionality through a variety of program or software methods to be spatially distributed without having all the functionality of a regular GIS. Direct incorporation of a tailored GIS functionality into the modeling system has been termed "isolation modeling" by Park and Wagner (1997).

Process modeling in combination with GIS is an effective method for representing propagation phenomena and other physical processes in space and time. Because of the limitations outlined above, however, we should not let the standard methods that are available and convenient dictate our conceptualization of nature rather than remain merely our best available representation of it (Raper, 1999). It could be a mistake always to view nature as a system of pixels or discrete polygons, or to accept objects as isolated within themselves.

4.4   GIS and environmental modeling

GIS and process models have been wed in a number of disciplines besides ecology. In fact, such applications have been made on much larger scales in oceanographic, atmospheric, hydrologic, geomorphologic, epidemiological, and other branches of science. These application communities have been brought together in a series of relatively large conferences under the title *Geographic Information Systems and Environmental Modeling* (*GISEM*). These international research conferences are designed to improve spatial–temporal predictive modeling of processes, events, and phenomena for environmental problem solving (Clarke *et al.*, 2002). The results of these conferences have been published in a series of edited volumes that demonstrate a wide range of applications, some ecological. Many contributions in these volumes are related to the problems and models presented later in this book, both as examples of applications and as formal treatises on technical and philosophical issues. Perusal of these is recommended for those wishing to learn more about the application of process models to GIS systems (Goodchild *et al.*, 1993, 1996; NCGIA, 1996; Clarke *et al.*, 2002).

For an excellent primer on GIS, we suggest Clarke's 2001 book, *Getting Started with Geographic Information Systems*. For a more extensive treatment of GIS, we suggest *Geographic Information Systems and Science* (Longley *et al.*, 2001). For a more in-depth source on research issues in GIS, we recommend the two volume set by Longley *et al.* (1999) titled *Geographical Information Systems. Principles and Technical*

Issues as a relatively up-to-date source volumes on the field, as well as the earlier edition of the same book by Maguire *et al.* (1991).

GIS has been adapted by some subdisciplines of ecology, perhaps most notably in vegetation and landscape mapping and analysis, population modeling, population ecology, restoration ecology, landscape ecology, and conservation biology. Turner and Gardner (1991 a,b), and more recently Johnston (1998) have reviewed applications of GIS in ecology and GIS provides the operational platform for the models described in Part II of this book. We have organized the introduction of these models so that knowledge of GIS operations is not necessary. However, familiarity with GIS will deepen the reader's appreciation for the structure of the models and the meaning of the model outcomes, not to mention that it will be very difficult to do research of this type without GIS capability. For research applications of models like these, a deeper knowledge of GIS is recommended.

4.5   About "explicit" space

The terms "explicit space" or "spatially explicit models" are frequently used in landscape ecology and process modeling. What does this mean? A dictionary definition of explicit is "Fully and clearly expressed; leaving nothing merely implied; unequivocal; clearly developed or formulated" (Neufeldt and Guralnik, 1994). Kareiva and Wennergren (1995) state, "The phrase "spatially explicit" refers to the fact that the models keep track of the exact positions of plants and animals; consequently higher-level processes such as productivity or pestilence can be related to landscape geometries." Coincident with that definition, Farina (1997, p. 208) says ". . . [spatially explicit] models incorporate the complexity of the real-world landscape (topological and chorological components)." Also, "A spatially explicit model is structured in such a way that the precise location of each element (organism, population, habitat patch) is known in comparison to the landscape features (corridors, edges, woodlots, rivers, fields, forests, etc." However, Farina also remarks (1997, p. 209) that "The type of landscape used in these models [spatially explicit] can be real or artificial. In the first case few land uses or other characteristics are preferable. The artificial landscape is used to simulate the response of species."

From these two sources it is clear that while models are linked with definite, well-formulated spatial databases, these spatial databases may or may not represent an actual environment or landscape. "Explicit spatial data" includes either artificial or synthesized spatial data used for theoretical modeling.

In Part II, we mostly use spatial data derived from actual environments. In those cases, we will refer to the data as explicit–realistic, remembering that spatial data, no matter how carefully prepared, are only a model of the true world they attempt to represent. In cases in which we use artificial spatial data, we will refer to those data as explicit–artificial.

4.6   Summary

Propagation phenomena can be reasonably represented through process modeling on a GIS database platform. General theory for such phenomena exists and can be used to understand and generalize these processes. This GIS-based modeling approach actually weds two types of model: models of the environment and models of the process being represented. As these are models, they are abstractions of the phenomenon and the environment, reality itself. The spatial databases and the models are limited in their abilities to represent nature and, even if they appear to be realistic, we must not allow the data or the model structures to become realities for us. These structural considerations are apart from the more basic problem of the data quality of the represented environmental domain. Outputs can be no better than the inputs. Spatial data can contain a variety of errors that are just as damaging to distributed model outputs as the models themselves. For a description of such errors see Goodchild *et al.* (1993).

GIS is a powerful tool for doing many things, including implementation of process models in explicit space. We must remember, though, that it is limiting in its intellectual structure and in terms of the software available to implement the desired task. Nevertheless, GIS is an enabling toolbox/database system/science. A minimal set of characteristics of GIS is presented here to give the propagation models that follow more meaning to the reader.

What are termed propagation models here are termed transport models in other disciplines and contexts. There is a rich literature for such models in non-ecological disciplines and, to a very large extent, the models presented in this book are adaptations of approaches and models in the literature of those disciplines. Part II will introduce these approaches and models into an ecological point of view. To the extent possible, the models that follow will operate in an explicit–realistic spatial context. We do this for two reasons. First, ecologists must learn to work outside of an "ideal" abstracted framework and attempt to operate in a more realistic framework in order to help society to make better decisions for environmental management. Second, new insights may be gained with realistically portrayed environments that may not emerge with artificial spatial arrays.

PART II

5

Introduction to Part II

Relativity teaches us the connection between the different descriptions of one and the same reality.

Albert Einstein

5.1   Background

Ecology, like most scientific fields, suffers from our inability as busy humans to keep abreast of conceptual and technological advances in other disciplines. This is most true among seemingly unrelated endeavors, but it is also a problem within the environmental sciences in general and even within ecology itself. The problem is exacerbated by the current pace of progress. As Alvin Toffler (1970) expressed it in *Future Shock*, "The illiterate of the 21st century will not be those who cannot read and write, but those who cannot learn, unlearn, and relearn." While more highly trained scientists focus their considerable energy onto specialized topics, our ability to monitor and understand their work and to put it into a larger context is easily overwhelmed.

Despite this, we all recognize that much is gained from increasing our awareness of the strategies and advances of other scientists in other disciplines. This has fueled a trend in recent years away from individuals pursuing lonely courses towards "Eureka!" moments (e.g. Clements, 1916; Tansley, 1935) to projects built around multidisciplinary teams and collaborative efforts (e.g. Kittel *et al.*, 1995; Schimel *et al.*, 2000; IPCC, 2001a,b). Such efforts have enabled us to advance our understanding of complex earth systems, like the climate system, that would not have been possible had atmospheric scientists, oceanographers, ecologists, geologists, and others worked on pieces of the puzzle in isolation. Complex integrated systems require both detailed knowledge of these pieces and broad understanding of linkages and synergies.

We posit that the propagation of ecological influences is one way of conceptualizing ecological systems (Reiners and Driese, 2001). Going beyond such conceptualizations, as explored in Part I of this book, has, by and large, forced us to borrow methods from other disciplines to illustrate how propagations work and how they can be represented in dynamic models of transport processes. Our hope is that the following chapters will provide ecologists with information for understanding a broad range of strategies being used and developed to study transport in different settings. The overviews to follow are intended to stimulate ideas about their application to ecological problems and to provide starting points for more detailed exploration.

5.2   Overview of Part II

The remainder of the book is organized into eight chapters, each of which is devoted to a particular transport vector. While not covering all mechanisms driving transport in ecological systems, the chapters touch on key processes operating in terrestrial, aquatic, and atmospheric settings, including diffusion, colluvial transport, wind, fire, fluvial transport, animal movement, light, and sound in that order. Each chapter begins with a general description of the transport vector and the ecological settings in which it is important, followed by a discussion of the physical principles important for understanding how transport is accomplished by each vector. Then we review general modeling strategies used by researchers in relevant fields to simulate transport by the vector of interest. Specific examples for each vector illustrate successful types of modeling, and readers are provided with key citations leading to more detailed discussions of each topic. While the modeling surveys are not exhaustive, they highlight ways of thinking about simulation modeling, especially in spatially explicit contexts.

While reading these chapters, it is instructive to remain mindful of both the similarities and differences in the approaches of workers in the various disciplines. In the face of a wide range of physical processes, environmental settings, scales, and objectives, simulations models often converge on similar approaches. Such convergence suggests that new problems in ecology might benefit from these approaches, and that it is not always necessary to begin from scratch. Likewise, the differences in approaches specific to particular disciplines or environmental settings suggest ways of framing and solving problems that can serve as heuristic models. The purpose of the remainder of the book is not to provide comprehensive reviews of each of the eight scientific disciplines associated with the respective vectors but, instead, to provide general insights from across these disciplines and to stimulate multidisciplinary interaction.

5.3   Example models

A CD-ROM is included with the book and contains, for each of the transport vectors described in Chapters 6–8, an interactive example of a model developed to simulate transport by that vector. In most cases, these models were adapted with permission from researchers in the relevant fields and simplified to allow users to experiment with model sensitivity to changes in key input variables. In a few cases (e.g. light, animal movement) we developed simple models ourselves. All of the models require that the reader have access to ESRI's ArcView 3.2x running in an MS Windows environment. The CD includes help files for each model and the models are described in detail in the associated book chapters.

6

Diffusion

> Of course, one must not forget that analogies are no better than analogies, models nothing more than models, and hypotheses simply hypotheses.
>
> Akira Okubo (Okubo and Levin, 2001)

6.1   Transport system description

The term diffusion is used in different settings to apply to a variety of transport processes. In hydrology and atmospheric science, for example, diffusion refers to the dispersion of gases or suspended substances in water or air (e.g. Harris, 1979) (Fig. 6.1). In ecology, diffusion is used to describe the movement or spread of plant species and individual animals or populations through a landscape (e.g. Pastor *et al.*, 1998; Turchin, 1998; Okubo and Levin, 2001) (Fig. 6.1). These usages have one thing in common: there is movement of some entity from areas of high to low concentration. Choy and Reible (2000) noted that modelers often lump processes that are difficult to quantify (such as dispersion in porous media) into factors analogous to diffusion coefficients, sidestepping the necessity for mathematical descriptions of the physical process. We can distinguish, however, between situations where the transport vector is diffusion in its strict sense from situations where the diffusion equation is used to *model* a process by analogy, or where a diffusion coefficient is used to lump complex phenomena.

To enlarge upon these usages, atmospheric scientists refer to several types of diffusion, including eddy diffusion, dispersion, and molecular diffusion (Hemond and Fechner-Levy, 2000). In fact, eddy diffusion and dispersion depend on the mass flow of air or water (Fig. 6.1). These transport processes are modeled using diffusion equations, but the vector for transport is wind or the movement of water (see Chs. 8 and 10). In part, evoking the diffusion analogy for complex, heterogeneous mass flow processes is practical when the scale of flux is large. For example, the complexities of eddy diffusion can be simplified over time and space to resemble molecular diffusion. The same is true in ecology when diffusion is used to describe the spread of an invasive species of plant or animal. Organism dispersal actually depends on transport by wind, water,

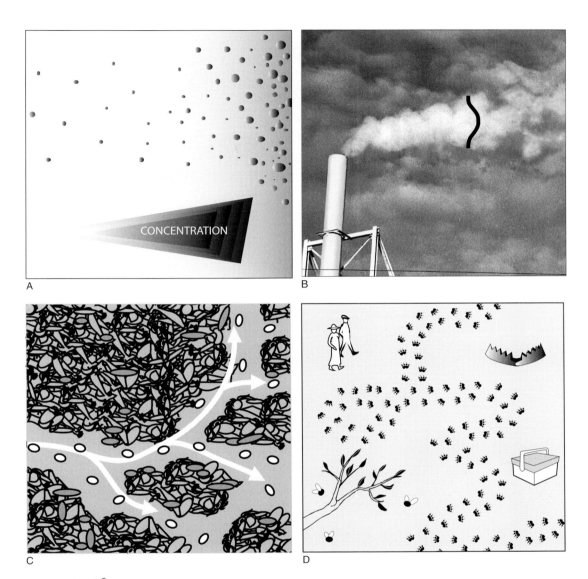

FIGURE 6.1
The concept of diffusion is used in ecology to describe very different transport processes across a variety of scales. In this chapter we limit our discussion to pure molecular diffusion (A) but diffusion can be used to describe many situations when random movement leads to transport. Diffusion caused by mixing of eddies in fluid flow is called turbulent diffusion (B) and is discussed in Chapters 8 (wind transport) and 10 (fluvial transport). Dispersion (C) occurs when particles are mixed as they move around solid obstructions in porous media, such as soil. Animal movement is often treated as a "random walk" (D) and modeled as analogous to diffusion.

A

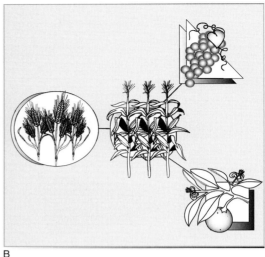

B

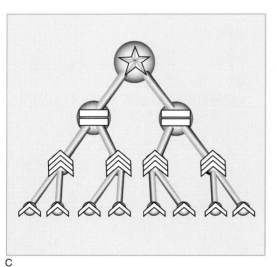

C

FIGURE 6.2

Non-molecular diffusion can be grouped into three conceptual types (Hagerstrand, 1967). Contagious diffusion requires direct contact as an effect is passed from one contiguous place to another (A). For stimulus diffusion to occur, a stimulus to a node causes that node to generate a stimulus of its own, which can stimulate other nodes; the spread of different types of agriculture occurs by this mechanism (B). In hierarchical diffusion (C), each node stimulates subordinate nodes (illustrated here with military ranks), which, in turn, stimulate nodes subordinate to them.

or animal locomotion rather than molecular diffusion processes, but viewed at a sufficiently broad temporal/spatial scale, the process may be represented as diffusion by analogy (Turchin, 1998).

Usage of the term diffusion is confused further by still other extensions of its meaning. Hagerstrand (1967) noted that there are three kinds of non-molecular diffusion: contagious, stimulus, and hierarchical (Fig. 6.2). "Contagious diffusion takes place as an effect is passed from one point to another contiguous point, that is, it depends on direct contact." This is exactly analogous to molecular diffusion. "Stimulus diffusion occurs when a process stimulates a node and the node stimulates other nodes by a stimulus of its own rather than the original stimulus." Stimulus diffusion is often described in a cultural context.

An example is nomadic Siberian tribes domesticating reindeer after hearing of domestication of horses by other tribes further south (Hoogvelt, 1976). "Hierarchical diffusion occurs when sensitive nodes are stimulated and they, in turn, stimulate a subordinate set of nodes that are in close proximity to the sensitive nodes and they, in turn, stimulate a set of nodes subordinate to them, *etc.*" All of these types of diffusion have utility in ecology, but we will only consider pure molecular diffusion in this chapter.

Because we treat mass flow (Ch. 7) and animal locomotion (Ch. 11) separately, we restrict our discussion of diffusion to its basic sense: molecular or particulate flux along concentration gradients driven by random movements of very small particles. Discussion of diffusion as an analogy for transport caused by other vectors is covered in more depth in the chapters describing those vectors. This narrow definition of diffusion restricts this transport vector to relatively small extents ranging from millimeters to meters, and to fluid environments without mass flow dynamics. One might question, therefore, whether diffusion plays any important ecological role. In fact, diffusion is pervasive over virtually all organism and environmental surfaces within boundary layers (e.g. soil–air boundaries, soil–stream boundaries and leaf–air boundaries) below the level at which transport by mass flow processes dominate. The ubiquity of diffusive fluxes means that they are important controls on ecosystem function and hence landscape structure. Many entities where transport by vectors like wind is ecologically important must first traverse thin boundary layers by molecular diffusion (Isard and Gage, 2001).

6.2   Underlying principles for mode of transport

The most basic mathematical expression of diffusion in a single dimension recognizes that diffusive movement is in response to a driving force (usually a concentration gradient) and is impeded by resistance. This relationship is captured in Fick's first law:

$$F_j(x) = -D_j(d\rho_j/dx) \tag{6.1}$$

where $F_j(x)$ is the gas flux density, D_j is the diffusivity of component j, and $d\rho_j/dx$ is the change in concentration relative to distance (the concentration gradient) (Campbell and Norman, 1998). The negative sign indicates flux from areas of high to low concentration. From this equation resistance can be approximated as:

$$\text{Resistance} \approx \Delta x/D. \tag{6.2}$$

While expressed above in one dimension, the model can be extended to *n*-dimensions using vector notation. The model can also be extended to describe diffusion from objects of various shapes (see Campbell and Norman, 1998) and has been used by environmental biologists to describe, for example, oxygen flux to roots and water vapor flux from leaves. It should also be noted that temperature plays a critical role. Molecular diffusion is ultimately caused by the random motion of individual molecules, which in turn is related to heat. As temperature increases so does the rate of diffusion. Many of the principles of diffusive transport are illustrated best in the discussions of specific situations. To minimize repetition, we include these principles below in the descriptions of models for particular environmental circumstances.

6.3   Modeling techniques

Models of transport by molecular diffusion can be classified into those that operate in liquid (usually water), gas (usually the atmosphere), porous solid media (e.g. snow or soil), and across the interfaces between these (e.g. the ocean–atmosphere boundary). The fundamental principles driving these models are similar, involving the movement of molecules, ions, or small particles along concentration gradients but mediated by resistance (Eq. 6.2).

6.3.1 Diffusion in gaseous media

Diffusion of gases or microscopic particles through the atmosphere is often overwhelmed by the presence of wind. Molecular diffusion occurs even in the wind but is relatively unimportant compared with turbulent transport and dispersion. The last two are discussed more fully in Ch. 8 and include models of smoke plumes, pheromone transport, pollen dispersion, and transport of airborne nutrients. The models of pure molecular diffusion discussed here are restricted to short-distance transport, such as occurs across surface boundary layers or embedded bulk flow of fluids.

Pheromones
Futrelle (1984) modeled the mechanism by which female silk moths (*Bombyx mori*) transmit mating pheromones to males of the same species. Male silk moths are capable of detecting these pheromones at extremely low concentrations and typically turn upwind when they encounter a pheromone plume downwind of the female moth. At the scale of silk moth chemoreceptors, pure molecular diffusion becomes important. Male moths detect the attractant, called bombykol, with tapered sensory hairs about 2 μm in diameter and 10 nm long. At a microscopic scale, bombykol molecules bounce erratically as they impact air molecules, a process that results in net movement of bombykol along its

FIGURE 6.3
The erratic random movement of pheromone molecules diffusing down concentration gradients increases the area that they traverse, as demonstrated in this schematic. Male moths have evolved to maximize their chances of intercepting a molecule by optimizing hair and pore spacing.

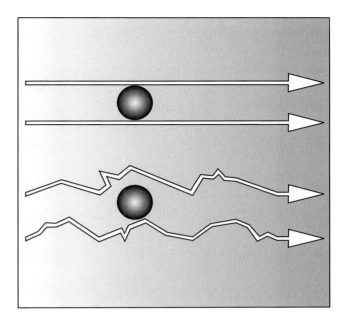

concentration gradient. The configuration of the male receptor hairs and pores maximizes the probability that these randomly moving pheromone molecules will be detected (Fig. 6.3). Futrelle modeled this molecular motion using probability distributions for molecular bouncing (similar to a "random walk"; Ch. 11) to trace pathways of individual molecules in space. He determined that much of the efficiency of male moths in detecting female pheromones results from the increased chance of the pheromone molecules hitting the receptor cells during diffusion. The random motion of bombykol molecules increased the effective diameter of the receptor hairs, and the pattern of receptor pores on the hairs optimized the chances of a pheromone molecule encountering a pore. Ultimately, all of these processes increased the chance of the male silk moth reproducing.

Plant leaves

Another example of molecular diffusion in the gas phase is the simultaneous loss of water vapor and intake of carbon dioxide (CO_2) through leaf stomata on plants. Stomata are pores on the surface of leaves that allow exchange of gases between the atmosphere and the interior of the leaf, where photosynthesis occurs (Fig. 6.4). They open and close in response to environmental conditions to provide CO_2 to the photosynthetic machinery (chloroplasts)

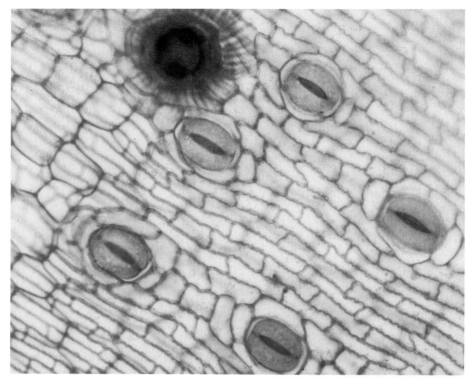

FIGURE 6.4

A magnified view of stomata (the eye-like features) on a species of *Catopsis* from the Bromeliad family (Bromeliaceae). Water vapor moves from the leaf interior to the atmosphere by diffusing along a concentration gradient created by the difference in concentration between saturated air inside the leaf and unsaturated air outside. Similarly, carbon dioxide can diffuse into the leaf, where it is used in photosynthesis. Resistance to diffusion is caused by the leaf boundary layer and characteristics of stomata as they respond to environmental factors. The blue ciliated structure at the top of the photograph is a scale unique to some Bromeliads. It helps the plant to absorb water and nutrients, much as roots do in non-epiphytes. (Photograph courtesy of G. K. Brown.)

within leaves. While stomata are open to let in CO_2, water vapor diffuses out.

Because the fixation of carbon by plants and the accompanying transpiration of water are important mediators of latent and sensible heat flux, leaf models form important components of land-to-atmosphere transfer schemes (e.g. Sellers *et al.* 1986; Dickinson *et al.* 1993) that are linked to climate models. Models of diffusion through stomata follow Fick's first law (Eq. 6.1), with flux driven by a concentration gradient and mediated by resistance. In this case, the photosynthetic process reduces the concentration of CO_2 within the leaf relative to that in the atmosphere. This creates the concentration gradient that drives

CO_2 uptake by diffusion. Similarly, the interior of the leaf is saturated with water vapor relative to the atmosphere, and the resulting water vapor gradient drives the transpiration of water outward through the stomata.

Resistance to flux from or into leaves is caused by a combination of several physical resistance factors. Bugbee (1985) described total resistance to CO_2 flux at the leaf–atmosphere boundary as:

$$r_{total} = r_a + r_s + r_m \qquad (6.3)$$

where r_a is boundary layer resistance, r_s is stomatal pore resistance, and r_m is resistance to diffusion encountered as CO_2 passes through the leaf mesophyll to the chloroplast where photosynthesis occurs. Alternatively, Dickinson *et al.* (1993) described stomatal resistance to water loss as a minimum stomatal resistance mediated by various environmental factors that partially control stoma opening and closing:

$$r_s = r_{smin} R_f S_f M_f V_f \qquad (6.4)$$

where R_f is the dependence of stomatal resistance on solar radiation, S_f is a seasonal temperature factor, M_f is a soil moisture factor and V_f is the vapor pressure deficit. Because all of these factors are greater than one, they all act to increase stomatal resistance.

Other modelers (e.g. Jarvis and McNaughton, 1986) have used the inverse of resistance, called stomatal conductance, to model diffusion through stomata. Stomatal conductance depends on the structure and function of individual stomata and their arrangement and density with regard to leaf area. Environmental factors can increase or decrease conductance as the stomata open and close in response to, for example, water stress.

All of these simple mathematical models are essentially one dimensional, with space considered only along linear concentration gradients. The size of the concentration gradient depends on the distance between an area of high concentration and one of low concentration, although this distance can be effectively lengthened in porous media by a property called **tortuosity** – the path length that a diffusing particle must trace to avoid obstructions (see the discussion of gas diffusion through snow below). In fact, extending these models to two dimensions by considering horizontal diffusion is important for many of these systems. Scaling diffusion models from individual stoma to entire leaves, to collections of leaves and finally to regional or global vegetation is a research area that is receiving considerable attention (e.g. Schulze *et al.*, 1994). Extending diffusion models from one dimension to two or three dimensions is especially critical for representing propagation into realistic environmental domains.

Pollutants

Contaminants in the atmosphere have profound effects on the health of living organisms and ecosystem function (Hemond and Fechner-Levy, 2000). While transport in air is accomplished over long distances primarily by advection, contaminants are subject to molecular diffusion while in transport (Choy and Reible, 2000). The net movement by molecular diffusion occurs at much slower rates than that in turbulent mixing. Choy and Reible (2000) noted that additional processes are also critical, including reaction of the contaminants with other chemical species, spontaneous degradation of contaminants, and dispersion.

Pollution transport modeling incorporates the physics of turbulent mixing, molecular diffusion, and other processes. Choy and Reible (2000) emphasized that models of contaminant transport must recognize the possibility of phase change during transport, include both advective and diffusion processes, describe reactions of the contaminants when they occur, and be applicable to a variety of geometries and boundary conditions. We refer readers to the Choy and Reible (2000) text for a more comprehensive discussion of these approaches. Other surveys of atmospheric pollutant modeling include Harris (1979), De Wispelaere (1981), Turner (1994), and Hemond and Fechner-Levy (2000).

6.3.2 Diffusion in aqueous media

According to Freeze and Cherry (1979), diffusion in solutions "is the process whereby ionic or molecular constituents move under the influence of their kinetic activity in the direction of their concentration gradient." Diffusion in water, like diffusion in the atmosphere, is complicated by the overwhelming influence of turbulent transport and dispersion. Freeze and Cherry noted that even in systems dominated by mass flow, like streams, diffusion can contribute to the overall mixing process, although diffusion coefficients in water are about five orders of magnitude less than in air or vapor. Mass flow in water is discussed in Ch. 10. We limit our discussion here to molecular diffusion in aqueous media. Examples include diffusion of materials into and out of individual cells and diffusion of nutrients, salts, and contaminants through water bodies.

Cellular diffusion

Diffusion at the cellular level in plants and animals is important for moving dissolved solids and gases across cell membranes and throughout protoplasmic volumes. Simple diffusion is responsible for the transport of such entities as water vapor, CO_2, oxygen, and urea into and out of cells. For simple diffusion to occur, these entities must be soluble in the lipid bilayer of the membrane or be small enough or energetic enough to pass through the membrane without

FIGURE 6.5
The zone from which materials can diffuse from soil pores into plant roots is called the diffusion shell. This shell represents the soil volume from which nutrients and other materials are available to a plant.

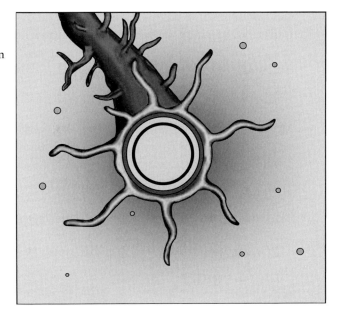

interference. Entities that do not possess these properties require energy inputs by the organism (active transport) for transport across cell membranes (Brown and Cooper, 1995; Campbell *et al.* 1999). Transport at the cellular level is discussed in depth in plant and animal physiological literature (e.g. Campbell *et al.* 1999) and we refer readers there. Suffice to say that this transport operates over short distances and is modeled using the Fickian transport equation (Eq. 6.1).

Cellular diffusion is ecologically important at the interface between a plant or animal and its environment and can create strong concentration gradients of transported entities. An example is the uptake of nutrients from the soil by plant roots. Nutrients in soil water move toward roots along concentration gradients created by root nutrient uptake, which lowers the nutrient concentration in a "diffusion shell" around the root (Nye and Tinke, 1977) (Fig. 6.5). Mineralization in the soil, which increases the concentration of available nutrients outside the diffusion shell, strengthens the gradient (Chapin, 1980). The availability of nutrients is influenced by the cation-exchange capacity of the soil itself, which can effectively remove nutrient species from the pool, diminishing the concentration gradient and the rate of diffusion for adsorbed nutrients. As a result, the thickness of the diffusion shell around roots varies for different nutrient species. Root density must be higher for plants to exploit nutrients that diffuse slowly (through thinner diffusion shells) than for those that diffuse more quickly. Chapin (1980) noted that soil particle size and moisture content determine the path length followed by diffusing ions. Soils with high moisture and small particle size allow nutrients to follow a more direct path to the root, steepening the concentration gradient and increasing the rate of diffusion.

Diffusion in lakes and streams

Modeling of plumes of toxins in oceans, lakes, and rivers is a robust area of research because of the unfortunate frequency with which spills of harmful substances occur. Much of the movement of toxins in these situations is associated with macroscale currents and turbulence in water bodies, but, just as we saw for atmospheric transport, movement resulting from molecular diffusion is imbedded in these bulk fluid flows. This causes plumes to widen with distance from the source, increasing the potential area of influence of the plume.

Hemond and Fechner-Levy (2000) and Schnoor (1996) discuss modeling of chemicals in surface waters, including streams, lakes, estuaries, and wetlands. The models that they describe, however, apply the Fickian diffusion equation (Eq. 6.1) to transport by turbulence or dispersion, and they do not generally consider pure molecular diffusion in these systems, perhaps because it is less important in defining the zone of influence of chemical contaminants. An edited survey of turbulent diffusion modeling by Harris (1979) also included descriptions of mixing models that used Fickian equations to simulate turbulent mixing in water.

6.3.3 Diffusion in porous media

Diffusion in porous media follows the same general principles as diffusion in the atmosphere or in water bodies, but the diffusion coefficients are typically much reduced because entities must take longer paths to avoid solid particles and because some of the transported entities are adsorbed (Freeze and Cherry, 1979). Because gas diffusion through water is four or five orders of magnitude slower than diffusion through gas-filled pore space in the soil, water in pores reduces the effective pore space available for diffusion even further (Campbell, 1985). Freeze and Cherry define an apparent diffusion coefficient, D^*, as:

$$D^* = D\omega \qquad (6.5)$$

where ω is an empirical value accounting for the effect of the solid phase of the porous media through which the entity is diffusing. The reader will notice that ω could also be modeled from physical characteristics of the porous media (i.e. porosity or tortuosity), which themselves are usually measured empirically. The apparent diffusion coefficient is substituted for D in the Fickian equation.

Gaseous diffusion in soils

Oxygen consumed and CO_2 respired by microorganisms in the soil and water vapor and other gases all diffuse through pore spaces in response to concentration gradients (Campbell, 1985). Diffusion of gases in soil follows Fick's law (Eq. 6.1), but in a porous medium, diffusivity is usually expressed as a product of the diffusivity of the gas in air and some function of air-filled porosity (Freeze and Cherry, 1979). This modified diffusivity is equivalent to the apparent diffusivity described above. Water-filled soils have lower porosity and higher tortuosity than relatively drier soils, and gas diffusion is slower as a consequence. Webb and Ho (1998) suggested that in some situations diffusion of a condensable vapor can be enhanced rather than slowed by the presence of liquid in the pores. They postulated that the mechanism for this may be that liquid acts as a site for local condensation and evaporation as well as increased temperature gradients in the gas phase compared with porous media without liquid. The original work on enhanced gas diffusion through porous media was done by Philip and de Vries (1957).

Campbell (1985) modeled oxygen diffusion in soil using the Fickian transport equation modified to account for porosity, as discussed above. He divided the soil into layers and assumed that flux across each element is linearly related to the concentration gradient of oxygen created by differences in gas concentration at "nodes" representing the boundaries between soil layers. Oxygen concentration is adjusted at each layer boundary based on biological consumption. Similar models can be constructed for diffusion using different geometries. Microorganisms represented as points in the soil might be nodes for spherical diffusion, while roots might cause cylindrical diffusion. Other authors (e.g. Livingston and Hutchinson, 1995; van Bochove *et al.* 1998) have discussed gas flux through soils using Fickian modeling. Hillel (1998) discussed the physics of gas diffusion in soils and provided mathematical descriptions.

Snow

Diffusion of gas from the soil surface upward through a snow pack to the free atmosphere can be an important source of greenhouse gas flux to the atmosphere in the winter months (Massman *et al.* 1997). We have chosen this phenomenon as an example of diffusion modeling, and a detailed discussion is included in its description below. Other authors have also treated diffusion of gases through winter snow packs (e.g. Brooks *et al.* 1995; Oechel *et al.* 1997; Mast *et al.* 1998; van Bochove *et al.* 1998, 2000) and they provide details of modeling strategies.

6.3.4 Diffusion across physical interfaces

Diffusion across the boundaries between media is arguably more important ecologically than diffusion within a single medium. Exchange between the ocean and the atmosphere, for example, has profound effects on biological and physical processes that operate at scales broader than that of the diffusion mediating the exchange (Tarrason *et al.* 1995; Sarmiento and LeQuere 1996; Lai and Patra 1998). Pools (like the ocean and the atmosphere) are often treated in models as sources or sinks of particular entities, with diffusion mediating the exchange. We cite examples here of diffusion across the ocean–atmosphere and soil–atmosphere interface.

Ocean–atmosphere boundary

The exchange of CO_2 between the ocean and the atmosphere is an example of transport across the interface between two physical systems that has important consequences globally, through its effect on the carbon cycle, and locally, through its effect on ocean biota and water chemistry (Sarmiento *et al.* 2000). This exchange is mediated by molecular diffusion driven by differences in the partial pressure of CO_2 (pCO_2) between the sea surface and the atmosphere. These pCO_2 differences create a concentration gradient that is the driving force for Fickian diffusion (Eq. 6.1). Resistance to diffusion arises from a combination of "air resistance" and "water resistance," each of which differs depending on the chemical species involved (Rodhe, 1992). Resistance in the water dominates for gases of low solubility (e.g. CO_2) that are unreactive in the aqueous phase. Air resistance dominates for other gases (Rodhe, 1992). These relationships are expressed in Henry's law, which describes the solubility of gases in water (Butcher, 1992).

Ocean and climate modelers recognize the importance of CO_2 exchange and have incorporated it implicitly into both ocean models and global circulation models (GCMs) (Archer *et al.* 2000; Sarmiento *et al.* 2000). Rather than explicitly treating diffusion at the molecular level, most of these broad-scale models base exchange rates on factors that control the concentration (pCO_2) differences that ultimately drive diffusion. Examples of these factors include sea surface temperature and ocean biological activity (and related nutrient levels), which, in turn, are controlled by broad-scale processes, like ocean circulation.

Archer *et al.* (2000) explored the sensitivity to diffusion of atmospheric CO_2 levels close to the ocean surface using simulation in two classes of ocean model. Box models treat the ocean as a set of internally well-mixed reservoirs. Ocean GCMs divide the ocean into a finite grid. Both types of model allow the consideration of spatial heterogeneity by dividing the sea into regions. Archer *et al.* (2000) found that explicit consideration of diffusion had strong effects on the modeled distribution of CO_2 from one region to another. These results highlight the importance of considering the factors driving diffusion in this system, since they have important global effects, even while operating only in thin boundary layers.

Soil–atmosphere boundary

Diffusion across the soil–atmosphere boundary layer, like that across the ocean–atmosphere interface, has far-reaching effects on climate and biota. An example is the flux of trace gases like nitrous oxide (N_2O) and nitric oxide (NO) from soils (Keller and Reiners, 1994; Liu *et al.* 1999; Davidson *et al.* 2000; Groffman *et al.* 2000). N_2O, a byproduct of nitrification and denitrification, is important partly because it is an effective greenhouse gas (300 times more effective than CO_2). NO contributes to tropospheric ozone production through photochemical reactions. Ozone has both climatic and chemical effects in the environment (Davidson *et al.* 2000).

Most models of soil–atmosphere flux treat diffusion implicitly by simulating the factors that control gas *production* and, hence, its concentration gradient. Firestone and Davidson (1989) described a conceptual "hole-in-the-pipe" (HIP) modeling approach where N_2O and NO leak from a pipe representing the flow of nitrogen through a soil ecosystem (see also Davidson and Verchot, 2000; Davidson *et al.* 2000). The sizes of the holes are determined by soil water content, pH, carbon, and the concentration of nitrogen oxides, all of which ultimately control N_2O and NO production (and microbial reduction) and soil diffusivity.

Soil water content is especially important because it controls both the amount of soil air space through which nitrogen oxides (and oxygen) diffuse and the anaerobic volume in soil aggregates where denitrification occurs. In Costa Rican forests, N_2O efflux increases exponentially with water-filled pore space (WFPS), probably because anaerobic production of N_2O in soil aggregates overwhelms the decreased soil diffusivity caused by pore water (Keller and Reiners, 1994). NO efflux, by comparison, is lower in wet forest soils where it is reduced by microbes faster than it can diffuse into the atmosphere. For both N_2O and NO, there is a complex relationship between gas production *within* the soil and diffusion *through* it. In Costa Rican forests, gas efflux is dominated by production, which determines the concentration gradient, rather than by the diffusion coefficient, a measure (the inverse) of Fickian resistance (Eq. 6.2). This relationship is further complicated by mass flow caused by wind or atmospheric conditions above the soil (Hillel, 1998).

Several simulation models empirically integrate nitrogen oxides production with diffusion to estimate efflux from soils into the atmosphere (Potter *et al.* 1996; Liu *et al.* 1999; Del Grosso *et al.* 2000; Plant 2000). Liu *et al.* (1999) modified the CENTURY model (Parton *et al.* 1987, 1994) to predict N_2O flux from tropical soils after their conversion from forest to pasture. Although CENTURY is more process based than the conceptual HIP model, diffusion is still largely derived from WFPS in the soil rather than directly calculated using the Fickian equation (Eq. 6.1). Reiners *et al.* (2002) used CENTURY with GIS data to explore the spatial and temporal variability in N_2O and NO emissions from Costa Rican soils but again treated diffusion implicitly, with rates controlled by gas production and WFPS. Whether diffusion is implicit or explicit, it is critical to consider both sides of the Fickian equation – the concentration gradient and the resistance – as well as the processes giving rise to them. In the case of nitrogen oxides, these are gas production and factors contributing to soil diffusivity, respectively.

6.4   Introduction to the example model

Snow can cover 44–53% of the land area of the northern hemisphere for more than half the year (Sommerfeld *et al.* 1993). Fluxes of greenhouse gases (e.g. CO_2, N_2O and methane) between the soil and the atmosphere were long assumed to cease beneath snow or when soil temperatures dropped to near 0 °C, but researchers have recently shown that the insulating effect of snow allows microbial and root respiration to continue in some environments (Brooks *et al.* 1995; Massman *et al.* 1997; Mast *et al.* 1998). These fluxes can represent an important part of the annual trace gas budget for some subalpine, alpine, and arctic ecosystems (Sommerfeld *et al.* 1993; Brooks *et al.* 1997). Even when extremely cold temperatures effectively stop microbial respiration, CO_2 fluxes are observed from some snow-covered arctic ecosystems, perhaps as a result of physical processes that release CO_2 early in the cold season (Oechel *et al.* 1997; Mast *et al.* 1998).

While the evidence for trace gas flux from snow-covered terrain is largely empirical, modeling offers a tool for exploring the effects of heterogeneous soil, snow, and meteorological environments. To demonstrate the modeling of gas diffusion through snow in a heterogeneous setting, we have modified a model called SNOWDIFF, originally formulated to simulate one-dimensional gas diffusion through the snow pack at one place on the landscape (Massman *et al.* 1997). By running this model iteratively, once for each cell in a tessellated (raster) landscape, and by allowing the model to use snow depths from a heterogeneous area, we can simulate the total CO_2 flux rate from the spatial domain.

6.4.1 Principles of the model

SNOWDIFF calculates the flux rate (mg/s per m^2) of gas from the top of a layered snow pack (Fig. 6.6) into the atmosphere using Fick's first law

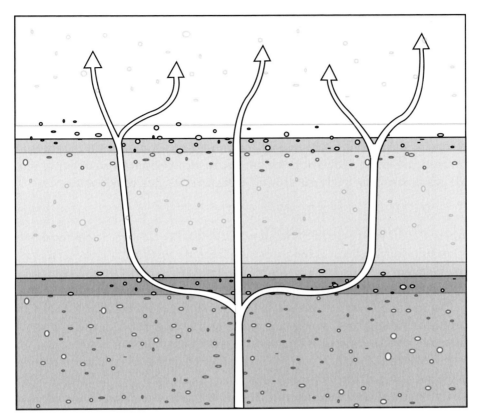

FIGURE 6.6
Gas diffusing through a layered snowpack from underlying soil. Layer properties,
including differences in porosity and tortuosity, affect the path followed by
individual gas molecules and the rate at which transport occurs.

(Eq. 6.1) modified to allow the snow pack to impede the gas flux (Massman
et al. 1995). Impedance (resistance) is determined in the model by snow depth,
number of snow layers, snow porosity, snow tortuosity, and snow temperature.
Porosity in this case is the ratio of empty pore space to total snow volume.
A tortuosity coefficient is used in the model to characterize the path length
that a molecule must traverse to move between two places in the snow pack.
Because the molecule must move around solid particles, this length is longer
in snow than it would be in the atmosphere. Tortuosity is difficult to measure
and is calculated as a function of porosity. Each layer in the snow pack can
have a different thickness, porosity, tortuosity, and temperature (Fig. 6.6). In
this way, the model recognizes that snow packs are vertically heterogeneous.
Values of these snow properties are based on field measurements from several sites (Massman *et al.* 1995) but model users can add their own measured
values.

The concentration gradient in the model is determined by the concentration of CO_2 at the snow/soil interface and the ambient CO_2 concentration at the top of the snow pack, both of which are based on *in situ* measurements (Massman *et al.* 1995). Again, these values can be varied by model users for particular sites. In the model, the driving forces, defined by this concentration gradient and the resistance to flux, created by the properties of the snow pack, are combined in a differential equation (Eq. 6.6) based on Fick's first law but with additional terms for gas movement caused by "pumping" resulting from differences in pressure above the snow pack. For pure diffusion without pressure pumping, Massman assumes steady state $(\partial \chi / \partial t = 0)$ and non-advection $(v = 0)$:

$$0 = (P_{00}/P_0)(T/T_0)(D_0/\eta)\partial/\partial z[(T/T_0)^{0.81}\eta\tau(\partial \chi /\partial z)] \qquad (6.6)$$

where T is the temperature of the snow pack $(^{\circ}C)$, $T_0 = 273.15\ ^{\circ}K$, z is the depth from snow surface (m), η is the porosity of layered snow pack, τ is the tortuosity coefficient of snow pack, χ is the mole fraction of CO_2 (ppmV), $P_{00} = 101.3$ kPa, P_0 is the ambient pressure, D_0 is the diffusivity of CO_2 in air at $0\ ^{\circ}C$ (0.139 m^2/s).

The model solves the equation using iterative numerical techniques until a solution is found that satisfies a lower boundary condition (at the soil–snow interface) within a specified tolerance (Massman *et al.* 1995).

The SNOWDIFF model calculates flux at one location using data describing snow pack properties. It is one dimensional in that it calculates diffusion only along a vertical line extending from the soil surface, through the snow pack, and into the atmosphere above. We extend the model to a heterogeneous domain by running it for each cell (5 m × 5 m) in a gridded landscape, but our modification is still one-dimensional since no horizontal diffusion between cells occurs and horizontal homogeneity is assumed within cells. For our version, we use the SNOW-TRAN model (Liston and Sturm, 1998; see Ch. 8) to generate snow depths across Libby Flats in the Medicine Bow Mountains of Wyoming. These depths are input to the SNOWDIFF and both snow depth and porosity can be adjusted by the user within realistic limits to explore model sensitivity to these properties. We assume homogeneous gas flux from the soil at the base of the snow pack. A future challenge will be to introduce spatial heterogeneity to the soil gas flux and snow properties to simulate better a realistic, heterogeneous pattern of diffusion across the domain. It may also be important to include horizontal transport *between* cells (Massman *et al.* 1995).

6.5   Summary

Diffusion operates across short distances but has profound consequences on ecological systems functioning at much broader scales, especially when diffusion occurs on vast scales such as along the ocean–atmosphere interface. Because fluxes across boundaries and between different physical systems (e.g. oceans and the atmosphere) are important transport mechanisms, and because molecular diffusion is a useful analogy for other types of transport, it is important to understand the mechanisms that control it. In this chapter, we have introduced the concept of Fick's law and cited examples of Fickian transport modeling in different settings. Readers interested in exploring this topic further are encouraged to experiment with the SNOWDIFF model provided on the CD and to investigate other models that are cited in the text. We emphasize that most diffusion modeling described in the literature is one dimensional and is not spatially explicit. Extension of diffusion models to heterogeneous environments is an area that needs further research, especially where consideration of horizontal transport is important.

Colluvial transport

No snowflake in the avalanche ever feels responsible.

Stanislaw Lec (Polish writer)

7.1 Transport system description

After years of repetitive freezing and thawing, a boulder the size of a small car breaks away from the top of a limestone cliff and falls, smashing into the ground 50 m below. The boulder gouges a sizeable hole in the soil at the cliff base before hurtling down the steep slope, destroying trees and shrubs, plowing an intermittent groove, and, finally, coming to rest on a terrace above a stream. Time passes. Plant communities differentiate between the north and the south side of the boulder in response to contrasting solar insolation. A pack rat nests in a rock hollow and forages among the vegetation in a wide radius around his nest. On the slope above, rainwater runs preferentially along the trail gouged by the boulder, forming a concentrated drainage channel. Relatively alkaline waters resulting from accelerated chemical weathering of the cliff scar flow down the new channel and enter the stream, raising the pH and, as a consequence, changing the invertebrate community near the confluence with the new drainage.

Colluvial transport or "mass movement" has profound geomorphic and eco-logic consequences. A single spatially and temporally isolated event, like the fall of the boulder described above, can affect plant and animal communities, biogeochemistry, and hydrology in significant ways in short and long terms, and at locations close to and distant from the initial event. Mass movements occur at a wide range of spatial and temporal scales and in very different envi-ronments (e.g. terrestrial or marine) with different effects. In this chapter, we discuss the phenomena of transport by mass movement in terms of initiat-ing events, transport mechanisms, and potential effects. We touch on some of the modeling techniques used to simulate colluvial transport and to predict the effects of different types of mass movement. Finally, as is true throughout this book, we ask readers to consider the network of interactions and types of transport that result from single events. In the falling boulder anecdote, colluvial transport led to changes in fluvial transport, light environments, and animal movement. Transport vectors seldom operate in isolation.

7.1.1 Classification of colluvial transport

We use the term colluvial transport synonymously with mass movement from the geomorphologic literature. Mass movement was defined by Summerfield (1991) as "the downslope movement of slope material under the influence of the gravitational force of the material itself and without the assistance of moving water, air or ice." Different authors have classified colluvial transport, mass movement, or mass wasting using various schemes (e.g. Sharpe, 1938; Varnes, 1958, 1975). Selby (1993) suggested that any classification should consider at least the velocity and mechanism of movement, the material being moved, the mode of deformation, the geometry of the moving mass, and the water content of the material. Summerfield (1991) offered a mechanistic classification using these criteria and we will adopt it here (Table 7.1). Mass movement is frequently labeled with the generic term "landslide", which groups together many types of specific event (e.g. debris slides, rock slides) and precludes others (e.g. rock fall). Instead we will use the specific terms applied by Summerfield.

7.2   Underlying principles for mode of transport

Ultimately, colluvial transport results from the force of gravity acting on materials, but circumstances can vary. Specifically, the types of mass movement (Table 7.1) are initiated by a variety of events or conditions; they transport different materials by different mechanisms; and they result in unique geomorphic, hydrological, and ecological effects. We discuss mass movement by dividing the discussion according to Summerfield's (1991) mechanistic classification of mass movement. Summerfield (1991) and Selby (1993) offer more detailed treatments.

7.2.1 Creep and heave

Creep is the slow, plastic deformation of rock, soil, or snow initiated by the stress imposed by the weight of overburden (Summerfield, 1991) (Fig. 7.1). Selby (1993) distinguished between creep, where individual soil particles move over exposed bedrock; depth creep, where soil or rock act as a deforming bed for movement of overburden; and soil creep, where soil moves under its own weight. Creep rates are generally in the range of 1 mm to 0.5 m per year. Donohue (1986) suggested four mechanisms responsible for creep on slopes: pure shear,

TABLE 7.1. *Scheme for classifying and describing important types of mass movement discussed in the text*

Primary mechanism	Mass movement type	Material type	Moisture content	Rate of movement
Creep	Rock creep	Rock (e.g. shale and clays)	Low	Very slow to extremely slow
	Continuous creep	Soil	Low	Very slow to extremely slow
Flows	Dry flow	Sand or silt	Very low	Rapid to extremely rapid
	Solifluction	Soil	High	Very slow to extremely slow
	Gelifluction	Soil	High	Very slow to extremely slow
	Mud flow	>80% clay-sized	Extremely high	Slow
	Slow earthflow	>80% clay-sized	Low	Slow
	Rapid earthflow	Soil containing clays	Very high	Very rapid
	Debris flow	Mixture of fine and coarse debris	High	Very rapid
	Debris avalanche	Rock debris (sometimes with ice and snow)	Low	Extremely rapid
	Snow avalanche	Snow and ice (sometimes with rock debris)	Low	Extremely rapid
	Slush avalanche	Water-saturated snow	Extremely high	Very rapid
Slides	Rock slide	Unfractured rock mass	Low	Very slow to extremely rapid
	Rock block slide	Fractured rock	Low	Moderate
	Debris/earth slide	Rock debris or soil	Low/moderate	Very slow to rapid
	Debris/earth block slide	Rock debris or soil	Low/moderate	Slow
	Rock slump	Rock	Low	Extremely slow to moderate
	Debris/earth slump	Rock debris or soil	Moderate	Slow
Heave	Soil creep	Soil	Low	Extremely slow
	Talus creep	Rock debris	Low	Extremely slow
Falls	Rock fall	Rock	Low	Extremely rapid
	Debris/earth fall	Cohesive soil units	Low	Very rapid
Subsidence	Cavity collapse	Rock or soil	Low	Very rapid
	Settlement	Soil	Low	Slow

Based on Summerfield (1991; Table 7.5) which was in turn based on Varnes (1978).

FIGURE 7.1
Creep can cause objects imbedded in the soil to bend or tilt downhill. Trees, utility poles, and gravestones are common examples.

viscous laminar flow, expansion and contraction (e.g. freeze–thaw), and particle diffusion. Effects of creep include bending of tree trunks, downslope curvature of strata near the surface, formation of small terraces (terracettes), and initiation of other types of mass movement (e.g. slides).

Heave, though distinguished from creep by Summerfield (1991), is really a special case of creep. Heave occurs when cycles of expansion and contraction cause downhill migration of rock (usually talus) or soil. Upon expansion, material is lifted normal to the slope; upon contraction, it settles in response to gravity (Fig. 7.2). The net result is slow downhill movement. Expansion and contraction at the surface is usually a result of wetting and drying, freezing and thawing, temperature changes, or animal burrowing (Summerfield, 1991). Rates of movement resulting from heave vary, depending on the magnitude of the expansion and contraction.

7.2.2 Flow

Flows are mass movements of soil, rock, or snow where mixing occurs throughout a moving mass without well-defined shear planes (Table 7.1 and Fig. 7.3) (Summerfield, 1991). Movement within the flowing mass is turbulent, though the flows themselves have distinct boundaries. Many types of flow occur, depending on substrate, water content, movement mechanisms, and movement rates. Summerfield (1991) included dry flows, solifluction, gelifluction, mud flows, slow and rapid earthflows, debris flows, debris avalanches,

Heave

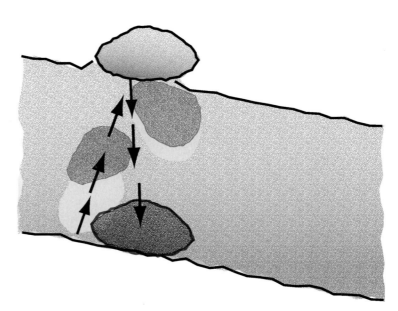

FIGURE 7.2
An idealized cross-section of material undergoing heave. Freezing or swelling of soil lifts material along a perpendicular to the slope, and then thawing or shrinking allows material to settle with gravity, causing net downslope movement. (Modified from Summerfield, 1991; Fig. 7.5.)

Flow

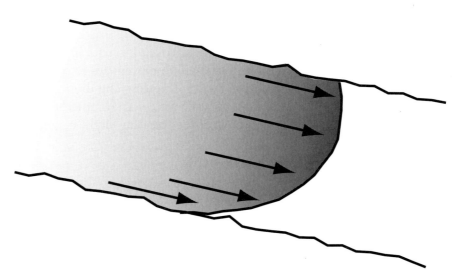

FIGURE 7.3
An idealized cross-section showing downslope flow. Material closest to the failure surface has lower net velocity than material higher in the flow. (Modified from Summerfield, 1991; Fig. 7.5.)

snow avalanches, and slush avalanches in a classification of types of flow. Selby (1993) advocated classification on the basis of rheological (mechanisms of flow) terms to reduce ambiguity. Most generally, flows are characterized by turbulent motion within viscous media, in contrast to slides (see below), which move as coherent masses along shear planes.

Frequently, flows are initiated by other types of mass movement; for example, slides or falls become flows when the moving materials break up (Summerfield, 1991). With the exception of relatively dry sand or loess flows (dry flows), most flows are mobilized by the addition of substantial amounts of water either from rain, snowmelt, glacial runoff, or lake overflows. It is the combination of factors such as the magnitude and duration of water supply, soil water status, relief, sediment availability, and others that ultimately determines thresholds for flow initiation.

Flow velocities vary from very slow, in the case of solifluction or gelifluction, to extremely fast, in the case of some debris or snow avalanches. Solifluction occupies the fuzzy area between creep and flow and occurs on shallow slopes (down to $1°$) at slow rates. At the other end of the spectrum, debris or snow avalanches can move with terrifying speed. A 1970 debris avalanche in Peru raced downhill at close to 300 km/h before becoming airborne as it leapt over a small ridge; it destroyed the towns of Yungay and Ranrahirca, which were in its path, and killed tens of thousands of people (Summerfield, 1991) (Fig. 7.4).

Snow avalanches deserve special mention because of their importance in alpine terrain, where they occur repetitiously along well-defined paths, often removing woody vegetation (Fig. 7.5). Like other flows, snow avalanches often begin as slides, with failure occurring along planes of weakness within the snow-pack (McClung and Schaerer, 1993). As a snow slab moves downhill, it breaks up, and motion becomes turbulent. Avalanches take many forms (e.g. dry slabs, wet slabs, ice avalanches, and slush avalanches) depending on the condition of the snow and the nature and orientation of the terrain – both in terms of steepness and the relationship to wind and sun. Effects of snow avalanches include the patterning of vegetation along avalanche paths (e.g. destruction of trees) and the downslope transport of rock, soil, and vegetative debris (McClung and Schaerer, 1993). Luckman (1971) attributed snow avalanches in the Canadian Rockies to the formation of many talus fields, which are themselves subject to creep, heave, and flow.

7.2.3 Slides

Slides are differentiated from other mass movements by the existence of well-defined shear surfaces (Table 7.1, Fig. 7.6). Summerfield (1991) describes two major types of slides based on the geometry of this surface, with planar shear

FIGURE 7.4
An aerial photo of the massive landslide that originated on Huascaran in the
Peruvian Andes and destroyed the town of Yungay in the foreground. (Photograph
courtesy of Servicio Aerofotografico Nacional de Peru, 13 June 1970.)

FIGURE 7.5
A snow avalanche path on Berthoud Pass, Colorado. Snowslides frequently prevent
the growth of trees and result in strong vegetation patterning in alpine areas. These
vegetation differences, in turn, affect a range of ecologically important processes,
such as hydrology and animal habitat use.

Slide

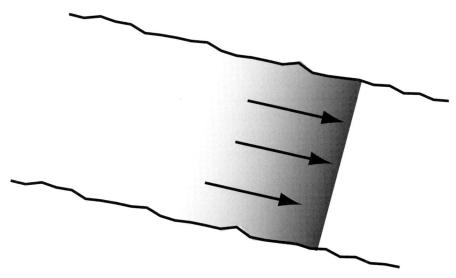

FIGURE 7.6
A schematic cross-section showing a slide. Material moves as a unit or block, sliding
along a definite shear plane. (Modified from Summerfield, 1991; Fig. 7.5.)

surfaces characterizing translational slides and concave surfaces characterizing rotational slides. While most slides are small and shallow, occurring on the scale of a few tens of meters in extent and 2–3 meters in depth, bedrock slides can be enormous, involving millions of cubic meters of material (Summerfield, 1999). The debris in large slides tends to break up; as a result the mass movement is transformed into flows.

Slide initiation requires the release of the entire material mass from a shear surface. According to Selby (1993), this can be accomplished either by factors contributing to high shear *stress* or by factors contributing to low shear *strength*. The former include removal of lateral support (e.g. erosion of toe slopes), increases in loading (e.g. addition of water weight), transitory stresses (e.g. earthquakes), removal of underlying support (e.g. mining), increases in lateral pressure (e.g. swelling clays), and increases in slope angle (e.g. tectonic tilting). The latter include the composition and texture of the materials, physicochemical reactions (e.g. clay hydration), effects of pore water (e.g. buoyancy effects), changes in structure (e.g. reorientation of clays caused by creep), changes in vegetation (e.g. logging or fires), and relict structures (e.g. old bedding planes) (Selby, 1993).

Once movement is initiated, either by the addition of shear stress or the reduction of shear strength, resistance diminishes rapidly and the mass moves downslope until resistance is increased again, usually by a reduction in slope angle at the base of the slide. In the case of rotational slides, which often occur in deep, homogeneous clays, the material commonly breaks into several segments by fissures that are oriented perpendicular to the direction of movement.

Many effects of slides are similar to those of flows – changes in hydrology, destruction of vegetation, reworking of landforms, etc. The Gros Ventre slide in northwestern Wyoming is an example. In the Gros Ventre mountains, fine-textured soils with high water-holding capacities have developed over sloping bedrock (Knight, 1994). On 23 June 1925, a block of earth about 2.4 km long, 0.8 km wide and up to 90 m deep slid and dammed the Gros Ventre River below, forming a lake, which still exists today (Fig. 7.7). A portion of the dam broke two years later, sending a pulse of water down the Gros Ventre, wiping out the town of Kelly, 3 km downstream, and eventually flooding the town of Wilson, 40 km downstream.

The Gros Ventre slide, which occurred in a relatively remote portion of the Gros Ventre mountains, had significant effects on the geomorphology of the slide path, the hydrology of the Gros Ventre River, the vegetation in the slide path, and the flood plain of the river for many kilometers downstream. It also caused damage to human communities along the river. In other words, a transport event in one part of the landscape at one time had profound effects across a larger spatial domain than that of the event itself, and over a long time period.

FIGURE 7.7
The Gros Ventre slide in the Gros Ventre Mountains near Jackson Hole, Wyoming.
The slide occurred in 1925 and created a lake when it dammed the Gros Ventre
River. Effects included immediate changes in the river and longer-term effects
resulting from both catastrophic failure of the original dam holding the new lake
and chronic changes in river hydrology.

7.2.4 Falls

Falls, like the boulder fall described at the beginning of this chapter, are the downward movement of rock, soil, or ice through the air and require the presence of very steep or overhanging slopes or cliffs (Table 7.1 and Fig. 7.8). Rock and ice are the most common participants in falls, because of their cliff-forming properties. Soil rarely occurs in locations steep enough to generate falls, although stream banks and bluffs are notable exceptions (Summerfield, 1991).

Initiating events for rock fall include physical weathering processes (e.g. freeze–thaw), the release of lateral pressure (e.g. erosion of overburden, leading to granite exfoliation), transitory events like earthquakes, and disturbance by animals (or humans). Soil falls usually occur on very steep banks that are eroding at their bases, either by the action of water or because of removal of material by humans and animals. Glacial movement in mountainous terrain is responsible for the majority of ice fall, but it can also be significant when water freezes to cliffs during periods of cold weather and subsequently breaks off when temperatures increase.

Falling rock, ice, and soil can have significant effects from their direct impact on the ground. For example, large trees do not survive at the bases of the enormous granite walls in Yosemite National Park, California, because they are obliterated by falling rock and ice (Fig. 7.8). Other effects are suggested by the falling rock anecdote at the beginning of this chapter: changes in hydrology, habitat, and even biogeochemistry are possible (Fig. 7.9).

7.2.5 Subsidence

Subsidence includes the collapse of material into underground cavities (e.g. caves) and more generalized settling of the earth's surface resulting from (i) pumping of oil, gas, or water; (ii) dissolution of limestone (sinkholes); (iii) collapse of underground mines; (iv) drainage of organic soils, or (v) wetting of dry soil (hydrocompaction). Effects of subsidence include changes in topography and hydrology, along with related impacts on plant and animal communities.

7.3 Modeling techniques

Because mass movements can have devastating effects on humans, much modeling in this field concentrates on landslide hazard prediction and slope stability analysis in steep terrain. Both of these topics are concerned more with the *potential* for mass movements than with movement or transport mechanisms themselves. Models typically estimate potential instabilities and then

FIGURE 7.8
The face of El Capitan in Yosemite National Park, California. Rock and ice falling
from large rock walls obliterates vegetation directly below and propagates effects
through space and time, as described in the falling-rock anecdote at the beginning
of Ch. 7.

FIGURE 7.9
A talus cone above Lake Louise in the Canadian Rockies. Obvious effects caused by rockfall, water flow pattern, and slope aspect are reflected by differences in vegetation on different parts of the cone.

incorporate various triggering factors. Notable exceptions include models of sediment transport and deposition into stream channels resulting from "landsliding" and debris flows (e.g. Benda and Cundy 1990, Benda and Dunne 1997).

Spatial heterogeneity is central to modeling colluvial transport because conditions leading to slope instability, including factors like slope angle, substrate characteristics, vegetation, and hydrology, are spatially variable. In many cases, the processes that lead to slope failure are reasonably well understood, but the spatial data needed accurately to parameterize spatial models do not exist and are difficult to obtain, even in the field.

For slides and other types of mass movement, spatially explicit modeling is ultimately dependent upon stochastic events (e.g. rainstorms or earthquakes). For this reason, quantifying transport by slides must be a statistical process, and explicit treatment of particular places at particular times is impossible. Spatially explicit models that describe slides after initiation are possible and may fall into any of the model classes described below.

7.3.1 Classification of models

A general classification scheme for mass movement models described by Selby (1993) is similar in many ways to the fluvial model classification by Refsgaard (1996), described in Ch. 10. We follow Selby's classification here and group models of hillslope processes into conceptual models, scale models, and mathematical models. A conceptual model is, according to Selby, "a mental image of a natural phenomenon in which the essential features of the phenomenon are retained and the details of which are regarded as extraneous [and] are excluded." Examples of conceptual models include maps or schematic diagrams explaining something about the phenomenon. Scale models are physical constructions of a phenomenon that are scaled with regard to materials and the stresses acting on them so that experiments can be performed on the model system. An example is the use of plaster blocks to examine the transfer of stress that occurs in real rocks.

The third major category of mass movement models described by Selby (1993) encompasses mathematical models. These are the most interesting for our purposes (transport in heterogeneous settings) because they have the potential to be both process based and spatially distributed. This potential is seldom realized, though, because of parameterization difficulties. Selby subdivided mathematical models into stochastic, statistical, and deterministic models. Stochastic models incorporate random processes describing the system or subsystem on the basis of probability. Examples include Markov chains and random walks. Statistical models also have random components, but these are naturally variable events like storms, earthquakes, or fires, where timing, magnitude, and location can be represented probabilistically. Deterministic models do not include random components and are often mathematical representations of conceptual models. Deterministic models may be process based or empirical. Because mass movement modeling involves complex processes and difficult-to-measure environmental parameters, few, if any, models are purely deterministic although many have deterministic components.

7.3.2 Hazard assessment

The potential for damage to roads and structures and loss of human life has generated a considerable body of theory and numerous models for the prediction and mapping of "landslide hazard" or slope stability. Hansen (1984) suggested that models of mass movement hazards have a higher chance of success than models of many other natural hazards, such as hurricanes, which are not "site-discriminant." Nevertheless, the stochastic nature of events and properties that initiate mass movement make time-specific predictive statements about specific sites impossible. According to Montgomery and Dietrich (1994), the five most widely used techniques for assessing hazards from mass movement include (i) field inspection; (ii) projection of future patterns of instability from analysis of existing slides; (iii) multivariate analysis of factors associated with existing slope instability; (iv) stability ranking based on criteria including slope, lithology, or geologic structure; and (v) failure probability analysis based on slope-stability models with stochastic hydrologic simulations. From the perspective of the discussion in this book, hazard assessment is an approach for modeling events or chronic conditions that *initiate* transport.

In the most general terms, mass movement is initiated when the shear stress along a potential failure surface exceeds the resistance to shear created by friction and cohesion (Carson, 1969). Shear stress is ultimately caused by the downslope component of the weight of the material. A mathematical expression for the threshold at which slope failure is initiated can be thought of as a "safety factor" (F) (Selby, 1993) for slope stability:

$$F = (\text{sum of resisting forces})/(\text{sum of driving forces}). \qquad (7.1)$$

When $F < 1$, the driving forces dominate and slopes are unstable. Similarly, when $F > 1$, resisting forces dominate. $F = 1$ is the threshold condition for slope instability. The science of estimating hazard potential or slope instability is really concerned with filling in the details of this equation. This includes development of equations or statistics that describe the resisting and driving forces, distributing these relationships across space and, finally, estimating the frequencies with which thresholds for slope instability might be exceeded. A discussion of the physics necessary to present explicit equations for the resisting and driving forces for each of the mass-movement types described in Table 7.1

is beyond the scope of this book. Brunsden and Prior (1984) and Selby (1993) have provided excellent general discussions of this topic. Hansen (1984) notes that

> ... one of the drawbacks of these methods is that parameters need to be carefully measured to derive meaningful results; however, measurement errors, sample disturbance, natural soil heterogeneity or limited knowledge of physical conditions of the slope can lead to uncertainty.

Sidle (1992) and Tang *et al.* (1997) have discussed the relationship between slope stability, initiation of mass movement, and tree roots. Tree roots add substantially to the resistive forces in the safety factor equation (Eq. 7.1) because they anchor soil in place. Stochastic events, like fire or logging, remove this resistive force and lower the threshold for slope instability, presenting an interesting modeling problem, which must include the temporal and spatial interaction of landscape pattern and slope processes.

Chung and Fabbri (1999) used a probabilistic approach for modeling landslide hazards with spatial data in a GIS. In their model, each pixel in the domain was assigned a joint conditional probability of being affected by a future landslide based on spatial input data layers for that pixel. Van Westen and Terlien (1996) used a more deterministic model for calculating safety factors for slopes in and near the city of Manzales, Colombia. They use the one-dimensional infinite slope model (presented by Graham (1984)) in a GIS environment to allow calculation of slope stability for each raster cell in a spatial domain. The model considered input maps depicting slope angle, soil depth, soil strength, and depth to groundwater. These calculations allowed estimation of susceptibility to shallow translational slides. Van Westen and Terlien (1996) suggested that one-dimensional models of this type are most suitable for use with GIS. Two- and three-dimensional methods are too difficult to parameterize.

7.3.3 Modeling transport by mass movement

Models that describe the actual movement of material occurring during colluvial transport are far less common than models assessing the initiation potential ("hazard potential") for these events. Examples of models simulating classes of mass movement from Summerfield (1991) are presented here to illustrate modeling approaches.

Models of creep
The rate of soil creep on a slope given a constant slope angle (β) is expressed by the equation:

$$C = K \sin \beta, \tag{7.2}$$

where K is a constant. This is an empirical, deterministic model of soil creep on slopes of uniform steepness in the sense that the value of K must be determined in the field and that output will be the same for identical input (β) values. Selby (1993), however, argued that such simple models neglect process and consequently are limited. Soil creep results from more complex forces than just the downhill component of gravity acting against a resisting force. While some of these complexities are incorporated empirically in K, some are not linear or require more mechanistic expressions.

A more process-based description of depth creep is based on Bingham rheological behavior and the equation:

$$\tau = \tau_0 + \eta\varepsilon, \tag{7.3}$$

where τ is shear stress, τ_0 is yield strength (creep threshold), η is a flow threshold, and ε is a shear strain rate (Selby, 1993). The elements of Eq. 7.3 can be further expanded based on process and soil properties, but the point here is that process-based expressions, *if they can be parameterized effectively*, add robustness to models.

Models of flow

Flows are usually initiated by events that add considerable water to slope materials, although dry flows are possible when, for example, sand exceeds its angle of repose on dune faces. Flows travel downhill, exhibiting many properties of turbulent fluids while moving large quantities of sediment, debris, or snow. Although much flow modeling concerns flow hazard prediction (see above), some researchers have simulated transport by flows with the aim of predicting, for example, their contribution to sediment loads in streams (e.g. Benda and Dunne 1997).

Johnson (1996) provided an account of the development of mathematical models describing granular movement within debris flows and slurries. He cited the work of Bagnold (e.g. 1954), who explored the phenomenon of "grain flow," in which the resistance to flow is caused by the impact of grains in the flow with one another. Johnson developed alternative "macro-viscous rheological models" to improve the description of the behavior of debris flows. Later, Johnson (1996) combined aspects of this work with some of the approaches of Bagnold to provide a more unified model. Iverson (1997) also provided a discussion of the physics of debris flows.

A number of models address the issue of sediment transport related to debris flows (e.g. Benda and Cundy, 1990; Benda and Dunne, 1997; see discussion of the example model below). Benda and Dunne (1997) used a stochastic model to interpret the temporal and spatial factors controlling sediment input to channels in a watershed in the US Pacific northwest. The model ran over a time domain of thousands of years (in annual time steps) and introduced stochastic fire and rainstorm events into the watershed. Source areas for debris flows were spatially distributed, but spatial prediction was not explicit, in the sense that properties of the source areas were provided by statistical distributions (describing slope gradient and colluvium depth). Fires reduce root strength in the source areas, decreasing resistance to slides.

Di Gregorio *et al.* (1999) used cellular automata to simulate a debris flow on Mount Ontake in Japan. They proposed their model as an alternative to deterministic approaches using differential equations and suggested that the latter "can involve almost insuperable or rough results when the simulations are based on differential equation solution methods, since it is extremely difficult to solve the governing flow equations." In their model, the domain is divided into grid cells, each of which is associated with altitude and the physical characteristics of the debris column resting on it. The "state" of each cell evolves over time depending on the states of the surrounding cells according to a transition function describing the physical process of the debris flow. In effect, this two-dimensional model transfers the debris slide downhill from cell to cell. The speed of the slide is determined by driving forces but is tempered by cell properties that contribute to frictional resistance. Other researchers (e.g. Barca *et al.*, 1986; Deangeli *et al.*, 1994) have also used the cellular automata approach.

Models of snow avalanches are mostly restricted to hazard identification (see above) and prediction of runout distances. When possible, such predictions are based on historical observation; however, when such data do not exist, researchers have used statistical and process-based approaches. According to McClung and Shaerer (1993), process-based simulations typically use "speed models" in which friction coefficients and slope angles are combined. Runout distance is determined by the slope at which the calculated avalanche speed reaches zero. Statistical approaches use probability distributions constructed from runout distances and terrain variables collected in specific mountain ranges comparable to the area of interest. McClung and Shaerer (1993) noted that both process-based and statistical methods have weaknesses, in the former owing to our lack of knowledge of the mechanical character of flows, and in the latter because of a lack of data.

Models of slides

As is the case for flows, much slide modeling is concerned with stability analysis of slopes to predict the *potential* for sliding. However, because translational slides usually occur along structural surfaces, like bedding planes in rock, initiation often depends upon nearly unmeasureable geological conditions (Selby, 1993). Rotational slides, in contrast, are generally initiated in relatively homogeneous clays, and stability analyses can be done using the same methods as for soils. In all cases, initiation depends upon stochastic events. Brunsden and Prior (1984) and Selby (1993) have discussed these issues.

Models of slides *after* they are in motion are mostly concerned with predicting the extent of the area affected by the slide: the length of the runout zone or the speed with which the slide moves. The simplest models for translational slides (along planar surfaces) treat them as a mass moving downhill, driven by the force of gravity and attenuated by friction between the slide mass and the shear plane. A coefficient of friction is usually measured empirically (e.g. Shreve, 1968). Another approach estimates the potential energy of the slide and uses this to predict probable runout. According to Peitersen (1994), both of these methods fail to account for important characteristics of slides, including the nature of materials in the slide, the shape of the slide path, whether the slide is constrained or unconstrained, and "lubrication" between the slide and the shear plane either by water or air. Because slides frequently evolve into flows, modeling may require consideration of both slide and flow mechanisms.

Models of falls

Modeling of falls, as we have seen for flows and slides, is mostly directed towards prediction of hazards to humans. Workers attempt to identify areas where falls are probable and to predict the potential extent of runout. Approaches include assessing evidence of previous falls (usually talus), with the assumption that new falls are most likely in these areas and will probably remain within the reach of the existing talus (Evans and Hungr, 1993; Wieczorek *et al.*, 1999). Estimating the potential energy of material in fall-initiation zones is a second approach for predicting falls that are likely to run beyond the existing talus. Wieczorek *et al.* (1999) used these methods in Yosemite National Park to predict hazards in this heavily visited, narrow valley bounded by rock walls. Tianchi (1983) used a mathematical model to predict rockfall extent.

Models of subsidence

Subsidence is the collapse of material into voids, which occurs across a wide range of scales from the compaction of soils to the breakdown of limestone caves. Modeling of these processes is well developed, especially in the realms of groundwater research (Poland, 1973; Freeze and Cherry, 1979) and extractive industries (e.g. Ren *et al.*, 1987). Modeling approaches include optimization strategies for choosing well locations and mathematical simulations that consider the relationship between the geometry of material that is removed (e.g. from mines) in relationship to the ground surface above (Freeze and Cherry, 1979; Gambolati *et al.*, 1991).

7.4   Introduction to the example model

A model simulating debris flows in mountainous terrain, based on work by Benda and Cundy (1990) for the Oregon Coast Range and the Washington Cascade Mountains (US), serves as the example model for this chapter (see the CD). Debris flows in these areas frequently reach confined mountain channels and travel down them for some distance, depositing material when channel slopes lessen or where tributary junction angles increase beyond empirically determined thresholds. While many of these flows are natural, studies demonstrate that their frequency is increased by human disturbances, including logging and road construction (Swanson and Lienkaempe, 1978). Flow deposition results in alteration of stream channels, which, in turn, affects hydrologic properties and fish habitat (Swanson *et al.* 1987b).

7.4.1 Principles of the model

The Benda and Cundy (1990) model is based on field observations of initiation, erosion, and deposition of debris flows in the mountains of the northwestern USA. These empirical data are used to predict travel distance and volume of debris deposition in mountain channels (Benda and Cundy, 1990). The model was designed to facilitate parameterization using terrain and channel characteristics obtainable in the field, air photographs, and elevation data. Importantly, the model does not explicitly consider complex rheological properties of the debris flows because of the difficulty (or impossibility) of parameterizing these properties. Implicit in this approach is the assumption that rheological properties of the flows are similar across the model domain and differences in flow behavior are explainable primarily by terrain characteristics across the model domain. Because of these assumptions, the model should be calibrated for use in other areas where rheological properties may differ. Recalibration is accomplished in the field by measuring gradients, tributary angles, and deposition volumes associated with particular topographic situations. These parameters are relatively easy to obtain; as a result, the model should be of practical use for land managers and scientists.

The core of the model structure determines where deposition occurs during flows by following a decision tree based on the empirical thresholds for debris deposition (Benda and Cundy, 1990; Fig. 7.10). Additional routines identify potential slide-initiation sites, trace stream channels, determine flow effects (e.g. scouring), and calculate characteristics of deposited debris (e.g. debris volume). The model version included with this chapter uses a raster (10 m × 10 m pixels) data structure with data layers for elevation and susceptibility to flow

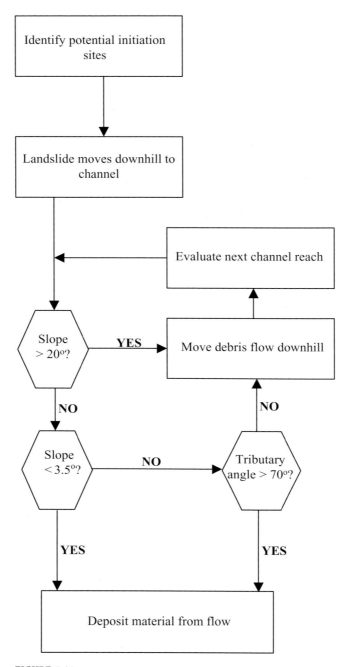

FIGURE 7.10
Decision tree showing the basic structure of the model. (Modified from Benda and Cundy, 1990.)

initiation. Flow initiation potential is based on a topographic index of slope stability that uses an algorithm described by Montgomery and Dietrich (1994). The algorithm considers surface gradient and the amount of drainage area per unit contour length for each pixel. During model runs, channels are traced using elevation data and channel-initiation points are calculated using slope, drainage area, and topographic convergence. Elevation data also allow determination of flow direction and channel gradient. Landslide source areas are assigned weights from the empirical probability of finding a slide in each pixel based on the calculated stability index.

Once the model has traced the channel network, determined flow direction, calculated channel gradients, and identified probable slide-initiation zones, potential debris flows are traced using the decision tree (Benda and Cundy, 1990; Fig. 7.10). Debris tracks are classified into scour, transport, depositional, and fan zones depending on channel gradient and tributary angles. Scouring results in accumulation of debris volume (an empirically determined volume per unit length of scoured channel). Subsequent deposition of the scoured material occurs when either channel gradient decreases below, or tributary intersection angle increases above, empirical thresholds. The volume of material deposited and the length of channel affected by deposition are determined by spreading the scoured volume down the receiving channel at a constant depth.

7.5   Summary

The rockfall event described at the beginning of this chapter illustrated how a relatively uncomplicated process (a boulder moving in response to gravity) has ecological consequences across space and through time. Other colluvial events, including creep, heave, flows, slides, falls, and subsidence, are more mechanistically complex and result in geomorphologic, hydrologic, and ecologic changes across scales. These events and processes are fundamentally heterogeneous in space and time, and modeling usually includes stochastic elements and/or complex parameterization, both of which make spatially explicit prediction difficult or impossible. At specific sites, though, modelers have used probabilistic techniques and process-based models to simulate both the potential hazard to humans and the behavior of the events themselves. Detailed discussions of colluvial transport are available in the geomorphologic literature and we recommend Summerfield (1991) and Selby (1993) for those interested in pursuing this topic further. We also include an example model based on the work of Benda and Cundy (1990), which uses some of the principles discussed in the text.

Wind transport

But besides creatures fully formed, the atmosphere contains innumerable
germs of future life, such as the eggs of insects and seeds of plants, the latter
provided with light hairy or feathery appendages, by means of which they are
wafted through the air during long autumnal wanderings. Even the fertilizing
dust or pollen from the anthers of the male flowers . . . is carried over land and
sea, by winds and by the agency of winged insects . . .

Alexander Von Humboldt (1850)

8.1   Transport system description

In the Sahel, drought parches the landscape and Harmattan dust rises
high into the atmosphere, moving westward over the Atlantic (Prospero *et al.*,
1981; Stoorvogel *et al.*, 1997). Dust particles settle into the ocean, providing iron
and phosphorus to phytoplankton in the nutrient-limited waters below (Fung
et al., 2000). Some of the dust remains airborne across the Atlantic, depositing
nutrients on the Antilles and *Aspergillus* sp. spores into shallow Caribbean waters
(Shinn *et al.*, 2000). These fungal spores drift downward onto coral reefs, ger-
minating and inoculating the sea fan species *Gorgonia ventalina* and *G. flabellum*,
eventually killing them and degrading coral reefs throughout the Caribbean
Sea. The link between local soil erosion in the Sahel and marine ecosystems in
the Caribbean is wind, the subject of this chapter.

Wind is the motion of atmospheric air relative to objects. At the global scale,
it ultimately results from the uneven distribution of solar radiation from low to
high latitudes (see Ch. 2). At regional scales, wind originates from the movement
of air from areas of high to low pressure in the atmosphere or in response
to contrasts between land and oceans (Green, 1999). High and low pressure
areas themselves are embedded into global scale circulation. At finer scales,
differences in topography, surface reflectance, vegetation, or individual clouds
can generate wind. At all of these scales, and in all geographic locations, wind is
critical to ecosystem function because it is a transport vector for mass, energy,
and information.

We discuss here the primary mechanisms by which wind transport occurs
and some of the modeling approaches used for understanding the movement

of specific wind-transported entities. Many authors have treated specific types of wind transport and provide details we can only touch on examples. Summerfield (1991) offered an excellent description of aeolian processes and the resulting landforms from a geomorphologic perspective. Forman (1995) discussed wind and wind transport in the context of ecological and landscape mosaics. Hemond and Fechner-Levy (2000) emphasized the role of wind across scales, especially in the transport of atmospheric contaminants. Isard and Gage (2001) provided a comprehensive discussion of aerobiology: the transport of organisms by the atmosphere. These authors introduced readers to the rich literature addressing wind transport.

8.2   Underlying principles for mode of transport

Wind transport requires that entities first be detached from their original location and entrained in the wind, then carried horizontally and/or vertically through space and finally deposited at a location some distance from where the transport originated. This triad of processes – release and entrainment, horizontal and vertical transport, and sedimentation and deposition of entities – provides a framework for studying wind transport (Isard and Gage, 2001). Knowledge of these processes is important for understanding and modeling wind transport, and we organize our discussion around them.

8.2.1   Release and entrainment of entities

Entrainment of matter (e.g. soil, snow, pollen, water) into wind requires first that particles be small enough to be moved by wind and second that these particles be lifted or released from a surface. The weight of the individual particles, cohesion between particles, and their mechanical packing, electrostatic forces, and friction all act to resist entrainment. Movement is initiated when drag and lift forces, or the impact of other particles, overcome this resistance (Summerfield, 1991). Some biological particles, including seeds and pollen, are actively released into the wind by a variety of mechanisms (see p. 150), requiring energetic cost to the organism.

Drag is more important than lift in detaching and entraining particles (Summerfield, 1991). Drag is caused by differences in pressure between the windward and leeward sides of objects in wind. It can result in surface creep – the downwind rolling or sliding of particles – and it can move particles too large to be lifted. Drag force increases as the square of wind speed, so the amount of movement caused by drag depends strongly on wind velocity as well as on particle weight and cohesion between particles. Lift occurs when differences in air velocity above and below an object result in a pressure gradient. Particles

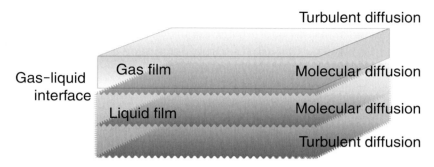

FIGURE 8.1
Molecular diffusion results in the movement of entities across boundary layers
between source or sink areas and regions where wind transport results in more rapid
transport.

experiencing lift can "jump" and become entrained in wind (Summerfield,
1991). The amount of lift is a function of the local wind speed, which itself
depends partly on the profile of the particle in relationship to the wind stream.
As we saw for drag, the potential for a particle to be lifted free of a surface
depends partly on its weight and partly on cohesion with other particles.

The direct impact of particles already entrained in wind with stationary particles is another mechanism for transport initiation. Airborne particles (usually
less than 0.1 mm in diameter) and larger entities (up to 0.4 mm or even larger)
experiencing saltation (bouncing transport) can transfer part or all of their
momentum to stationary objects. These direct impacts are important for initiating transport, especially of large particles (Summerfield, 1991).

Molecular diffusion (Ch. 6) is also important because it is the mechanism by
which entities traverse the boundary layer between source or sink zones and
flow regions, where longer distance transport occurs (Fig. 8.1). In this sense,
the diffusion vector contributes to the entrainment of entities that are then
transported over longer distances by wind. Diffusion occurs in all directions
in response to concentration gradients and is imbedded in broader-scale flow.
The spread of a gas or odor plume (see below) is partially caused by molecular
diffusion, although turbulent mixing (dispersion) is more important.

Finally, abrupt changes in wind speed (gustiness) can increase the rate at
which particles are entrained in the wind by overcoming the forces resisting
drag or lift (Aylor and Parlange, 1975; Aylor et al., 1981). Relatively high-energy
gusts initiate particle movement, which is then sustained by lower-energy
background wind.

8.2.2 Vertical and horizontal transport

In the atmosphere, vertical transport is accomplished by turbulent mixing, convection, and, at broader scales, by rise resulting from midlatitude and

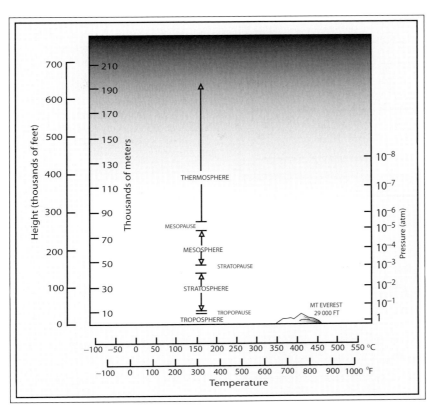

FIGURE 8.2
Atmospheric structure affects the rates of upward and downward vertical transport.
Residence times of entities entrained in the atmosphere are influenced as a result.

tropical disturbances (Hasse and Dobson, 1986). In addition, the vertical temperature profile and related atmospheric structuring (Fig. 8.2) affect rates of vertical transport and ultimately the residence times of transported entities. Each atmospheric layer may act as a relative source or a sink for particular entities. For example, nitrous oxide (N_2O) is chemically stable in the troposphere but decomposes photochemically in the stratosphere. Because mixing between these two layers is slow (Fig. 8.3), N_2O may exist for long periods and experience global transport in the troposphere before it is destroyed in the stratosphere (Rodhe, 1992).

Horizontal transport by wind depends on phenomena and mechanisms that operate across temporal and spatial scales (which are at least partly correlated; Rhode, 1992; Fig. 8.3). Hasse and Dobson (1986) and Rodhe (1992) have presented instructive discussions of horizontal transport in the context of time scales. From short (hours) to long periods (years), transport mechanisms range from advection by local winds (hours), diffusion by turbulence (hours), advection by synoptic and local wind (one day), advection in synoptic disturbance and

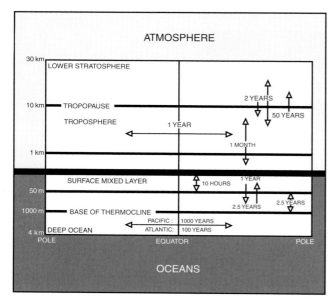

FIGURE 8.3
Horizontal and vertical transport in the atmosphere are affected by local and broad-scale heterogeneity. Transport across the equator between the northern and southern hemispheres is slower than transport within hemispheres. Similarly, transport between atmospheric layers is often slower than transport within layers.

tropospheric flow (days), tropospheric transport (months), stratospheric transport (years) and hemispheric exchange (years to decades) (Fig. 8.3). Residence times of specific entities in the atmosphere determine which of these transport mechanisms are meaningful for that entity.

Gravitational settling (sedimentation) in the atmosphere is also relevant because it partly controls the duration of entrainment (see below). Although sedimentation is a transport vector itself (see Ch. 2), it bears directly on wind transport. The rate of settling in fluids, including the atmosphere, can be expressed by Stokes' law. In its simplest form, Stokes' law applies to spherical particles settling at speeds that do not generate their own turbulence (Reynolds number < 1) (Hemond and Fechner-Levy, 2000):

$$w_f = ((2/9)gr^2 \Delta \rho)/\mu_f, \tag{8.1}$$

where w_f is the settling velocity (length/time), g is gravitational acceleration (length/time squared), r is the particle radius (length), $\Delta \rho$ is the difference between particle density and air density (mass/volume), and μ_f is the dynamic viscosity of air (mass (length multiplied by time)). For airborne particles of very small radii ($< 1 \mu m$) or low densities ($< 1 g/cm^3$), gravitational settling can usually be ignored (Hemond and Fechner-Levy, 2000).

In the remainder of this section, we discuss some of the mechanisms that are especially relevant to horizontal and vertical transport in the atmosphere, including dispersion and turbulent transport, advection, and the special case of vortices. Dispersion and turbulent transport dominate vertical movement, while advection is arguably more important horizontally. Clearly, the

FIGURE 8.4

Wind velocity increases logarithmically with height above the boundary layer over rough surfaces. This relationship is altered by stability conditions. Unstable atmospheric conditions "stretch" eddies upwards and stable conditions "flatten" eddies. Vertical transport in the atmosphere above rough surfaces is influenced by these relationships. (Modified from Thom, 1975.)

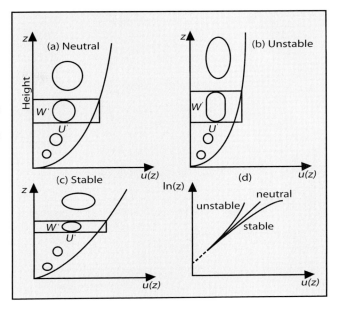

interaction between vertical and horizontal transport is fundamental to atmospheric transport.

Dispersion and turbulent transport

Eddy diffusion, dispersion, and turbulent transport are all caused by random mixing of turbulent eddies, which themselves result from the interaction of wind with a rough surface (mechanical turbulence) or from buoyancy (convective eddies). Modeling of turbulent transport is complex because of the chaotic nature of turbulence (Garratt, 1992; Driese and Reiners, 1998). In the most general sense, turbulence results in the transport of entities along concentration gradients; however, unlike in pure molecular diffusion (Ch. 6), the entities are mixed mechanically by local air movement (eddies) and the resulting fluxes are larger.

The wind-speed profile close to the ground but above a narrow zone known as the roughness sublayer (where complex pressure effects occur) is logarithmic and can be described by the equation (Thom, 1975):

$$u_z = (u^*/k)\ln[(z-d)/z_0] \tag{8.2}$$

where u_z is the wind speed at height z, u^* is the friction velocity (a measure of the wind velocity in eddies), k is von Karman's constant, d is the zero plane displacement height (the height at which momentum transfer between the wind and the surface occurs), and z_0 is the roughness length (an empirically derived surface characteristic) (Fig. 8.4). Modeling of turbulent transfer between a surface and

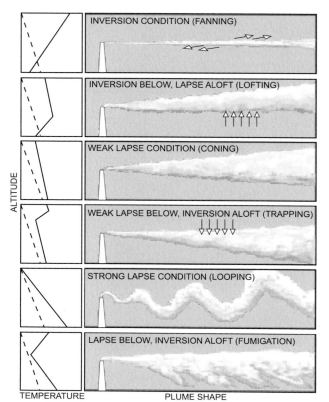

FIGURE 8.5
Patterns downwind of smokestacks illustrate the effects of atmospheric conditions on the path followed by windborne entities. (Modified from Hemond and Fechner-Levy, 2000; Fig. 4-10.)

the atmosphere is usually based on this equation, although parameter estimation is mathematically difficult (Driese and Reiners, 1998). Physical conditions, including issues resulting from differences in atmospheric stability (Fig. 8.4) or from the presence of the roughness sublayer (Raupach *et al.*, 1991), further complicate the process. Abtew *et al.* (1989), Kustas *et al.* (1989), Garratt (1992), Driese and Reiners (1998), and many others have provided details on these problems.

Buoyant turbulence is caused by heating or cooling of the air near the earth's surface (Turner, 1994). Heating by the sun, for example, can result in convective eddies and thermals that rise over 1000 m. Radiative cooling of the surface at night can result in inversions, which reduce vertical turbulent transport. A particularly noticeable effect of buoyancy is the shape and trajectory of smoke plumes released into the atmosphere as they move downwind of their source and undergo dispersion (Turner, 1994; Fig. 8.5).

Advection

At regional (synoptic) or global scales, transport may be more closely linked to the general trajectories of wind flow than to local winds and turbulence.

The jet stream, for example, wraps around the globe and quickly transports volcanic ash or embedded storm systems over long distances. Planetary waves caused by broad-scale turbulence transport warm air from near the surface to the upper atmosphere, a process that is expressed in the movement of storms from west to east across the globe (Berner and Berner, 1996). At regional scales, wind resulting from pressure gradients is an important transport vector. Advection at this scale transports entities like pollen, dust, and pollutants across continents. Predictive meteorology and the study of global climate are concerned with these types of transport (Kahl *et al.*, 1991; Kahl, 1996).

Vortices

Vortices (areas of air spinning around a vertical axis) occur across scales. Whirlwinds, just a meter or so in diameter, hurl dust and leaves upward 5 to 10 m into the air and then collapse. "Dust devils," visible because of entrained soil, are a common sight during summer over dry areas or fallow fields. These features are tens of meters in diameter and can carry dust hundreds of meters up into the air, where it may then be carried kilometers by advection. Even larger are tornadoes, whose funnels, dark with dust and debris, realize wind velocities of up to 140 m/s. Tornadoes can lead to transport of sizeable objects (e.g. sacks of flour, jars of pickles) for tens of kilometers and lighter materials (e.g. canceled checks, marriage certificates) for hundreds of kilometers (Snow *et al.*, 1995). Larger still are the vortices of hurricanes and typhoons that form over warm oceans; these carry enormous amounts of water, momentum, and latent energy for thousands of kilometers. Hurricanes and tropical storms are responsible for a significant portion of the precipitation received in the mid-Atlantic region of the USA. Finally, the pressure cells that pass through the temperate zones of Earth are very large vortices. These cells are the principal distributors of energy and water from the equatorial zones poleward (Riehl, 1978).

The circular or spiraling pattern of vortices is caused by the interaction of wind with the rotation of the Earth. In the northern hemisphere, cyclones occur when wind moving down pressure gradients towards a low-pressure center is diverted to the right by the Coriolis effect, causing the wind to spin into a counterclockwise rotation. Likewise, in high-pressure areas, wind moving outward from high to low pressure is diverted to the right, creating clockwise spin (Riehl, 1978). Tornadoes show a similar counterclockwise rotation in the northern hemisphere, suggesting that they are an extreme example of converging air interacting with the Earth's rotation. At the scale of dust devils, however, no such systematic rotation is seen. Notably, Riehl (1978) pointed out that the story of water spiraling down drains in opposite directions in the northern and southern hemispheres is a myth!

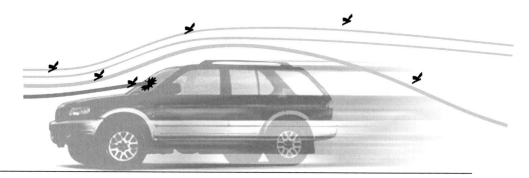

FIGURE 8.6
Particles entrained in wind follow streamlines and may, therefore, avoid impact with obstructions. (Modified from Hemond and Fechner-Levy, 2000; Fig. 4-30.)

8.2.3 Deposition and momentum transfer

The deposition of material, energy, and information represents the final step in the triad of processes that make up wind transport. Matter is removed from the atmosphere by dry or wet deposition, or it is altered by chemical transformations. In the last, material is not physically removed but is transformed into material with different properties. This may affect its residence time in the wind and, as a consequence, its potential for transport.

Dry deposition is the absorption or retention of airborne materials at a surface (i.e. soil or vegetation) (Eliassen 1980), ultimately resulting from sedimentation (see discussion in Section 8.2.2) and impaction. Impaction occurs when particles in the wind stream run into objects in their path. Insects smashed on a car windshield are unfortunate examples of impaction. Such impacts, though, do not *always* occur, because moving air flows around objects (along streamlines) and embedded entities tend to move with the air, avoiding impact (Fig. 8.6). The forces required to move objects along the streamlines are transferred via the fluid (air), and impaction occurs only when an object has more momentum than can be overcome by these forces (Hemond and Fechner-Levy, 2000). Many insects apparently in the path of disaster are actually carried over or around moving cars.

Wet deposition is the removal of atmospheric materials by rain or snow, or the interaction of cloud droplets (fog) with vegetation or the ground surface (Lovett and Reiners, 1986). Many atmospheric constituents, including some gases whose rates of sedimentation or impaction are very slow or close to zero, can be quickly removed from the air by these processes (Hemond and Fechner-Levy, 2000). Particulate matter, including pollen, spores, and dust grains, can also be washed out of the atmosphere by precipitation.

The third mechanism for "removal" of atmospheric constituents is their chemical conversion to other forms. An example is the oxidation of sulfur dioxide (SO_2) to sulfate in the atmosphere (Schlesinger, 1997). Both of these substances are soluble and are ultimately removed from the atmosphere in rain (wet deposition). The rate at which materials are altered in the atmosphere determines their residence time, a critical concept for understanding the extent of atmospheric transport (Murray, 1992). Entities with long residence times can potentially be transported farther by wind than those that remain in the atmosphere only for a short time (Hemond and Fechner-Levy, 2000).

The transfer of energy as momentum from the atmosphere to the earth's surface is apparent in the movement of vegetation in a breeze or the flapping of a flag. The momentum of moving air molecules is transferred to objects by friction, by the direct impact of air molecules, or by pressure effects (e.g. lift and drag). According to Thom (1975), the very presence of turbulence suggests that a surface is experiencing a downward flux of momentum, since turbulence is *caused* by a momentum transfer, called "drag." Thom emphasized that wind drag is "the primary flux associated with the frictionally driven eddies in turbulent boundary layer flow."

8.3   Modeling techniques

8.3.1 Types of wind transport model

Models of wind transport range from simple "scale" models, usually involving wind tunnels (e.g. Woodruff and Zingg, 1953), to complex, process-based simulation models (e.g. Liston and Sturm, 1998). Edmonds (1979), in his text on "aerobiology," classified atmospheric transport models within the framework of "system modeling." As Isard and Gage (2001) illustrated, system models span the gamut of complexity from simple box and arrow diagrams to complex computer algorithms. The common characteristic of model organization is that "related components . . . act together in time and space" (Isard and Gage, 2001). Within the framework of "system models," Burleigh *et al.* (1979) divided modeling approaches into deterministic simulation models, statistical models, and synoptic models. Stochastic approaches, although not mentioned by Burleigh *et al.* (1979), are applicable to certain components, for example when random wind gusts initiate the release and entrainment of pollen particles into the wind flow (Aylor, 1990; Jackson and Lyford, 1999).

8.3.2 Modeling wind transport of specific entities

Matter

Particulate matter in the atmosphere is defined as discrete solid or liquid particles ranging in size from a few nanometers to about 100 μm in diameter

(Hemond and Fechner-Levy, 2000). Particulate material with diameter $< 50\ \mu m$ is an aerosol. Aerosols are called primary when emitted directly into the atmosphere and secondary when they result from conversion of gases already in the air to solids or liquids. Most of the windborne solid and liquid material we discuss here, with the exception of seeds and organisms, are considered aerosols. Wind-blown dust, sea spray, volcanic emissions, and plant particles typically have diameters in the $10\ \mu m$ size range (Hemond and Fechner-Levy, 2000). Many droplets are smaller but are still considered aerosols. Wind-blown snow may be larger than aerosols depending on conditions. We organize our discussion here around specific entities, including particulates and gases, rather than by size classes, but we recognize that particle size is critical for determining sedimentation rates (Eq. 8.1) and residence times.

SOILS

Erosion and redeposition of soil and associated nutrients by wind is significant across spatial scales and in most ecological systems (Sterk and Stein, 1997). Much of the literature on this topic concerns agricultural erosion (e.g. Foster, 1991) and is motivated by the direct consequences for humans. At global scales, however, soil transport also has profound ecological consequences. As described at the beginning of this chapter, dust from the Sahel is deposited as far away as the Caribbean Sea (Prospero *et al.*, 1981), where it may cause damage to coral reefs (Shinn *et al.*, 2000). Chadwick *et al.* (1999) suggested that Hawaiian rainforests on highly weathered soils would be nutrient deprived were it not for atmospheric inputs from marine aerosols and phosphorus-rich dust from central Asia (more than 6000 km away). Atmospheric transport of soil and nutrients across such a broad range of scales requires modeling approaches tailored to specific phenomena. We touch on a few of these here, starting with local transport.

At local scales, wind transport modeling focuses largely on soil erosion, for the same reasons that soil erosion by water receives so much attention (see Ch. 10). Wind erosion in agricultural areas has significant negative consequences for humans (e.g. the "dust bowl" of the 1930s) and its prediction helps to direct farming practices (Foster, 1991). The principles on which agricultural wind-erosion modeling is based, however, are applicable to other systems and provide a foundation for the general treatment of soil transport by wind. Foster (1991) discussed wind-erosion modeling and highlights parallels to historic water-erosion modeling. The similarities are not surprising since both are specific instances of the general case of erosion by "fluid" flow over a surface.

A conceptually simple model, the empirical wind-erosion equation (WEQ; Woodruff and Sidoway, 1965), is analogous to the "Universal Soil Loss

Equation" (USLE; Ch. 10), which predicts soil erosion by water. The WEQ relates annual loss of soil (E) by wind to a set of empirically derived factors:

$$E = f(I, K, C, L, V), \tag{8.3}$$

where I is a soil erodability index, K is the soil surface roughness, C is a climatic factor related to wind speed and soil surface wetness, L is "fetch" (the length of the field across which wind can blow unobstructed), and V is a vegetative cover factor. Values for the factors are determined by field measurement, but in practice are often based on measurements made for specific locations in Kansas, where early development of this equation occurred.

Sterk and Stein (1997) adopted two geostatistical approaches, kriging and simulated annealing, to model empirical maps of sediment removal and deposition. The former is a smoothing technique that "provides a predictor with minimal variance of the prediction error at every position" but "does not reproduce the statistical properties of the observations" (Sterk and Stein, 1997). The latter uses conditional stochastic methods to maintain the statistical properties of the field data but does not provide the best possible predictions. Sterk and Stein (1997) found that maps based on kriging were useful for predicting soil loss, but not for spatial depiction of the erosional processes. Simulated annealing better depicted the spatial variability. Although these methods both have a spatial component, they are essentially empirical.

Empirical models, like the WEQ or the geostatistical techniques discussed above, are limited because they apply only to the situations from which they have parameters. Process-based models offer the opportunity to apply physical principles that can be distributed across heterogeneous systems. The necessity for consideration of heterogeneity in wind-erosion modeling led to the "Wind Erosion Prediction System" (WEPS) by the US Agricultural Research Service (Foster, 1991). WEPS capitalizes on earlier work on specific aspects of wind-erosion physics (e.g. Bagnold, 1941) and includes submodels for simulating weather, crop growth, decomposition, hydrology, soil properties, tillage practices, and erosion.

The transport of dust and nutrients over long distances depends primarily on broad wind patterns (e.g. Prospero et al., 1981; Kahl et al., 1991; Chadwick et al., 1999). At continental or global scales, advection is more important than local turbulence, and transport follows prevailing winds. Processes causing entrainment in source areas or deposition in sinks are usually not explicitly considered but instead are treated as average flux rates. Regional climate models, general circulation models (GCMs), and trajectory models (Kahl et al., 1991; Kahl, 1996) are useful tools for predicting dust transport at these scales (see p. 152).

SNOW TRANSPORT

Snow redistribution by wind has profound ecological effects in many regions. In open, windswept environments, including alpine tundra, arctic tundra, and grasslands and shrublands in cold climates, the redistribution of snow is a primary control on water availability and related ecological properties of the landscape (Knight, 1994; Driese *et al.*, 1997). In steep terrain, wind loading of snow increases the likelihood of avalanches on some slopes – a secondary effect of transport (see Ch. 7).

The demonstration model for this chapter, Snow Tran-3D (Liston and Sturm, 1998) is a spatially distributed snow transport model and we discuss it in detail in Section 8.4. Snow Tran-3D is instructive because it considers spatial heterogeneity in topography and vegetation. Other modelers (e.g. Pomeroy *et al.*, 1993) have examined the transport and sublimation of snow in process-based models, but for flat terrain and steady-state conditions. Kind (1981, 1986) discussed the physics and modeling of snow drifting for various surface-roughness conditions (e.g. snow fences, buildings, natural obstructions). Other researchers (e.g. Iverson 1980) have used wind tunnel experiments to examine snow drifting.

SALT SPRAY

The ocean is an enormous source of aerosols to the atmosphere. As bubbles near the sea surface burst, they fling small droplets of seawater into the air and these droplets can become entrained in wind. Eventually, the water evaporates, leaving solid particles of sea salt that mirror the composition of seawater (Schlesinger, 1997). Most sea-salt aerosols settle quickly back into the ocean but some remain in the atmosphere and are transported globally by wind, sometimes depositing salts on land (e.g. McDowell *et al.*, 1990) or contributing to atmospheric chemistry (Schlesinger, 1997). Modeling of sea-salt transport, like that of other particulates, depends on the scale of interest and is rarely spatially explicit.

CONTAMINANT TRANSPORT

One could hypothesize a hyperbolic relationship between the potential impact to humans and the number of models and modeling approaches. When human health or economic well-being is at issue, modeling is prolific. This is abundantly true for atmospheric contaminant transport, and many comprehensive discussions on this topic are found in the literature. Hemond and Fechner-Levy (2000) have given an excellent general treatment of the physics of the atmosphere relevant to the transport of chemical contaminants and the processes of transport and deposition. Harris (1979) and Turner (1994) have provided more detailed discussions of turbulent diffusion relating to transport. De Wispelaere

(1981), in a collection of papers from a conference on air pollution modeling, offered various approaches at different scales. We touch here on only a few of the most frequently applied modeling approaches for predicting the fate of atmospheric contaminants.

The diversity of atmospheric transport models is partly a result of the diversity of initiating events. Many contaminants are carried in the atmosphere, including gases (e.g. CO, CO_2, SO_x, NO_x, methane (CH_4)) and particulates (discrete solid or liquid particles) ranging in size from about 100 μm (e.g. volcanic ash, plant particles) to aerosols (< 50 μm) to ultrafine aerosols (< 0.1 μm) (Hemond and Fechner-Levy, 2000). These constituents are released from many sources by many mechanisms. Point sources, like smokestacks or volcanoes, are a classic case for which models have been developed; however, other sources, like wildfires or agricultural areas, can have broad geographic source areas.

Hemond and Fechner-Levy (2000) used scale as an organizing principle for discussing chemical transport in the atmosphere and we follow their discussion here. They recognized that "the spatial scales on which advective and Fickian [see Ch. 6] transport occur in the atmosphere span an extraordinarily large range, from a few tens of meters in indoor air to thousands of kilometers globally." They further emphasize that:

> If chemical transformations are relatively slow compared with mixing rates, it may be appropriate to use simple box models in which air volumes are considered to be well mixed (i.e. no concentration gradients exist). When chemicals are transported significant distances in the amount of time it takes for the relevant chemical reactions to reach equilibrium, the most useful models are commonly based on solutions to the advection–dispersion–reaction equation.

For different temporal and spatial scales, modeling must consider both the transport mechanism and the rate at which the transported entities are transformed chemically.

At local scales (~ 10 km), contaminant modeling frequently concerns the behavior of plumes originating from point sources like smokestacks. Plumes extend downwind of their source as a result of advection (bulk flow) while simultaneously spreading with turbulent mixing, which itself is usually modeled as an analogue to Fickian diffusion (Turner, 1994; Hemond and Fechner-Levy, 2000; Ch. 6). Both the advective trajectory followed by the plume and the rate of turbulent dispersion are influenced by atmospheric stability conditions (Fig. 8.5).

For short-range plume dispersion under steady-state conditions and homogeneous terrain, Gaussian plume models are frequently used. These assume that the distribution of contaminant concentration in a plume follows a Gaussian

FIGURE 8.7
Entities carried in plumes move downwind by advection while spreading through turbulent diffusion. Typically, the concentration of entrained material follows a Gaussian concentration gradient from the center of the plume outwards.

distribution, with the highest concentration in the center (the top of the "bell") and the tails of the curve spreading outwards (Fig. 8.7). "Gaussian puffs" have also been modeled using similar assumptions but for intermittent release of material into the air (De Wispelaere, 1981). According to Turner (1994), Gaussian plume modeling does not work well for convective situations or for releases at ground level. The strategy has been used for modeling transport of many entities in the atmosphere, including contaminants, pollen (Jackson and Lyford, 1999), and pheromone plumes (Elkinton et al., 1984).

At broader scales (tens to hundreds of square kilometers), multiple point sources are often "lumped" by assuming that their total emissions are mixed uniformly throughout the volume of air below a mixing height in the atmosphere (Hemond and Fechner-Levy, 2000). This type of "box model" works best when winds are light and the mixing volume of air is contained, as might be seen when an area is surrounded by hills (Fig. 8.8). Models of this type are often applied in urban settings to estimate air quality when emissions come from multiple sources (like automobiles).

For long-range transport of airborne contaminants (continental or intercontinental), models can estimate the concentration and deposition of contaminants on daily or shorter time intervals, or, more commonly, for longer averaging periods (Eliassen, 1980; Miller, 1987). At these scales, models consider both advection and mixing but require numerical techniques to solve complex and computer-intensive algorithms. Eliassen (1980) described Eulerian and

FIGURE 8.8
Air pollution trapped around the city of Monterrey, Mexico. Monterrey occupies a
valley surrounded by mountain ranges and hills.

Lagrangian numerical methods and the advantages of each. Models adopting
these approaches include the "Regional Acid Deposition Model" (RADM), the
"Acid Deposition and Oxidant Model" (ADOM), and the "Regional Lagrangian
Model of Air Pollution" (RELMAP). Miller (1987) reviewed trajectory models
that use meteorological data to examine past transport in the atmosphere. Trajectory models have also been used to simulate future paths of contaminants
(e.g. Kahl *et al.*, 1991; Kahl, 1996; Fig. 8.9).

Global-scale transport of materials with long residence times (e.g. CH_4, N_2O,
CO_2) is much more rapid within hemispheres than between hemispheres. Full
tropospheric mixing within either the northern or the southern hemispheres
requires only a few weeks, but mixing between hemispheres takes one to two
years (Hemond and Fechner-Levy, 2000). Similarly, mixing between the troposphere and the stratosphere requires 20 to 30 years and is thought to occur
mostly through tropical upwelling. Only materials with very long residence
times are subject to this scale of transport, and the mechanisms are still research areas. Modeling of global trophospheric transport is in the realm of
GCMs, which are spatially distributed using a coarse three-dimensional raster
(see p. 152).

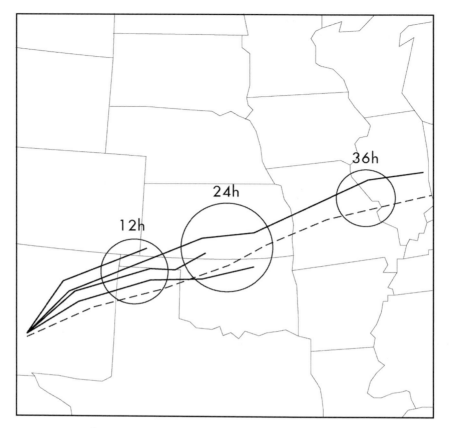

FIGURE 8.9
Trajectory modeling uses wind forecasts to predict paths followed by windborne contaminants released at specific points. The multiple paths shown represent predictions based on forecasts from different times preceding release. Forecasts closer to the release time are typically more accurate than those well in advance. Circles represent forecast areas affected by the contaminant at 12, 24, and 36 hours after release. (Modified from Kahl *et al.*, 1991.)

TRANSPORT OF ORGANISMS

It is impossible to pigeon-hole the airborne transport of organisms (even pigeons) into categories of matter, energy, or information since all three are associated with biota. Furthermore, potentially airborne organisms span a range of sizes from, for example, microscopic bacteria to Andean condors with wingspans of 4 m. The movement and transport of biota through the atmosphere is known most generally as aerobiology (Isard and Gage, 2001), though many disciplines (e.g. botany, ornithology, mycology) are concerned with particular aspects of the field. McAtee (1917) reported that "spores, pollen grains, algae, diatoms, infusoria, rotifers, living mites, mussels weighing up to 2 ounces, fish, salamanders, frogs, toads, turtles, rats and many organic fragments" have all been isolated from the air or have fallen out of it. While there are general physical principles, the relevance of particular transport mechanisms will change depending on the organism. The passive entrainment of a bacteria and the active takeoff of a condor both involve the physics of lift, but simulation modeling of the two organisms requires different approaches (see Ch. 11).

Although the distinction is fuzzy, we restrict our discussion in this chapter to relatively passive transport by wind of organisms like bacteria. While Andean condors may be transported against their will in violent storms, under normal conditions they have some control over the direction and distance that they travel. Even insects can, to a large degree, control their direction and rate of travel (Wolfenbarger, 1946), although they become passive travelers at lower wind speeds than do condors. We discuss active transport by animal movement in Ch. 11. Isard and Gage (2001) have provided a comprehensive discussion of the transport of living organisms and disseminules in the atmosphere, both by active and passive means, and we rely on their discussion here.

Edmonds' (1979) conceptual "aerobiology process model" included five phases relevant to modeling passive transport of airborne biota across scales. Isard and Gage (2001) suggested that these phases be called "source, takeoff and ascent, horizontal transport, descent and landing, and impact" to generalize them for organisms of different sizes. Ideal terms may vary for transport of different entities, but the Isard and Gage terminology is a useful starting point for modeling. The systems approach proposed by Edmonds (1979) is particularly well suited to airborne transport of biota. Burleigh *et al.* (1979) described a systems model by Forrester (1961) that predicted microbial input to and removal from the atmosphere from a variety of potential sources and sinks. Isard and Gage (2001) presented examples of systems models for describing the spread of blue mold (a tobacco pathogen), black cutworm moths, and western equine encephalitis (by mosquitoes).

Wolfenbarger (1946), in an extensive survey of dispersion (not always by wind) of "small organisms," including viruses, bacteria, fungi, plants, and insects, used a statistical approach to relate the total density of organisms to distance traveled from known source areas. These models were one dimensional in that only distance from the source (not direction) was considered. Carroll and Viglierchio (1981) studied the distances that soil nematodes might be carried downwind from different source areas using a mechanistic Gaussian plume model (see p. 147). Although essentially one dimensional, this mechanistic model considered different source configurations (point versus line). Additionally, its mechanistic structure would theoretically allow some spatial distribution, at least to the extent that heterogeneous source areas and wind fields could be parameterized.

Passive transport of insects and other organisms over long distances by wind is not uncommon (Pedgley, 1990). Biota in some cases are thought to position themselves to take advantage of broad-scale flow (Johnson, 1969; Turchin, 1998). On the scale of hundreds to thousands of kilometers, GCMs (see below) could be used to predict transport of organisms based on the assumption that they will move with air parcels (Turchin, 1998). At these scales, the bulk movement of air is more important than the mechanics of turbulence.

Energy

Wind transports energy either directly as the momentum of atmospheric constituents in motion or through its movement of latent and sensible heat. Modeling of atmospheric transport of these energetic entities falls within the realm of climate models. At the global scale, GCMs simulate the earth's climate system and the long-range transport that results from the redistribution of energy globally by the climate system. Regional scale (mesoscale) climate models have also been developed to improve incorporation of the effects of finer-scale land-surface heterogeneity on the climate system (e.g. Pielke *et al.*, 1996). Specialized models that explicitly treat the exchange of materials and energy between the land surface and the atmosphere are known collectively as land–atmosphere transfer schemes (LATS) or vegetation–atmosphere transfer schemes (VATS) (e.g. Dickinson *et al.*, 1986; Sellers *et al.*, 1986). All of these modeling strategies recognize the importance of wind transport at various scales.

GENERAL CIRCULATION MODELS

GCMs are three-dimensional, process-based representations of the earth's climate. Their development has been motivated partly by the specter of human-induced climate change resulting from the observed increases in "greenhouse" gases over the twentieth century. Sulzman *et al.* (1995) reviewed the structure, applicability, and limitations of GCMs for environmental managers. They suggested that GCMs are not so much predictive tools as a means to explore the sensitivities of climate to various forcings (initial conditions). Despite this, researchers and policy makers rely on GCMs at least partly to constrain future climate change scenarios (Lemonick, 2001). Furthermore, the range of predicted climates from a collection of GCMs provides strong consensus for decision making (IPCC, 2001a,b).

GCMs simulate the circulation of the atmosphere (including wind) using the principle of conservation of mass, momentum, and energy; the ideal gas law; and the physical relationships describing radiation, clouds, and precipitation (Sulzman *et al.*, 1995). They fit the definition of systems models because various parts of the global climate system (i.e., atmosphere, oceans, and land surface) are treated separately but linked to varying degrees. Because they represent a global domain, and because of the complexity of the calculations necessary to simulate the many interacting components of the Earth's climate system, GCMs operate at a coarse spatial resolution, typically ranging from 2.5° to 8° in latitude and 2.5° to 10° in longitude with 2–11 vertical layers (Sulzman *et al.*, 1995). Spatial heterogeneity within these large grid cells is usually either averaged or the properties of the dominant constituent (e.g. land cover) are applied to the entire cell. Oleson *et al.* (1997) explored the effects of land cover aggregating for climate modeling and found that for some cover types (e.g. wet types with high evaporation rates), retention of subgrid heterogeneity can significantly affect model output. Other researchers (e.g. Avissar, 1992; Koster and Suarez, 1992; Seth *et al.*, 1994) have also explored the effects of spatial heterogeneity in coarse-scale climate models and have recognized its importance.

REGIONAL CIRCULATION MODELS

The coarse spatial resolution of GCMs is not sufficient for simulating climate in areas where heterogeneity in surface features, like vegetation, terrain, or ocean circulation are important. Sulzman *et al.* (1995) stated that "While current GCMs realistically reproduce global patterns of temperature and precipitation, the limitations imposed by simplifications of topography, oceans, and precipitation-generating processes often distort climatology at the regional level." The climate (including advective wind) of the western USA, for example, is strongly influenced by the Rocky Mountains, which, at the scale of a GCM grid cell, are smoothed into obscurity. Regional or mesoscale models operating on finer grids are one solution to this problem. Examples include the Penn State/NCAR Mesoscale Model Version 4 (MM4) (Giorgi *et al.*, 1990) and the Colorado State University "Regional Atmospheric Modeling System" (RAMS) (Pielke *et al.*, 1992).

Regional models necessarily simulate climate for smaller domains than GCMs but GCMs are commonly used to provide boundary conditions to the regional models. The interaction between GCMs and regional models can be one-way, meaning that there is no feedback from the regional model back to the GCM, or two-way when feedbacks occur. Two-way models, in which regional models are nested within coarser GCM grid cells, are difficult and expensive to implement (Sulzman *et al.*, 1995). Regional models can be coupled to ecological or hydrological models (like RHESSys) (Pielke *et al.*, 1996) that simulate land-surface processes and account for spatial heterogeneity. These types of coupled models can also be one- or two-way.

LAND–ATMOSPHERE TRANSFER SCHEMES

In a description of the "Biosphere Atmosphere Transfer Scheme" (BATS) by Dickinson *et al.* (1986) the authors state that:

The purposes of the biosphere–atmosphere transfer scheme (BATS), as coupled with the NCAR Community Climate Model (CCM), are to (a) calculate the transfers of momentum, heat and moisture between the earth's surface and atmospheric layers, (b) determine values for wind, moisture, and temperature in the atmosphere, within vegetation canopies, and at the level of surface observations, and (c) to determine (over land and sea ice) values of temperature and moisture quantities at the earth's surface.

To accomplish this, climate modelers have developed process-based models that use an energy-balance approach and simple hydrology to calculate fluxes of momentum, sensible and latent heat, and material (mostly water) between the surface and the atmosphere. Wind contributes to these fluxes largely through turbulence resulting from its interaction with vegetation and the ground surface. Examples of LATS include BATS (Dickinson *et al.*, 1986), the "Simple Biosphere Model" (SiB) (Sellers *et al.*, 1986) and the "Land Surface Transfer Scheme" (LSX) (Thompson and Pollard, 1995). These models use grid-based representations of the land surface and typically are three-dimensional in the sense that they include multilayered soils.

Information

TRANSPORT OF GENETIC INFORMATION

A propagule is "a structure, such as a seed or spore, which gives rise to a new plant" (Harris and Harris, 1994). We will also include here the transport of haploid pollen, which contributes to the spread of plants through sexual reproduction, and entities including but not limited to bacteria, fungi, animal eggs, and larvae. Collectively these "disseminules" participate in the dispersal of living organisms. The principles of airborne transport that apply to all of them are similar, although mechanisms of release and the ultimate effect of their transport differ.

Movement of airborne disseminules can be conceptualized within the framework of Edmonds' (1979) "aerobiological process model" (see p. 150). In the context of pollen transport, Laursen (2001) suggested use of the modified terms "presentation, release, transport, sedimentation, and impact" to describe the five stages presented by Edmonds. Presentation is the stage of biological development at which disseminules become available for transport. In the case of pollen, for example, presentation occurs when the anther dehisces, exposing the pollen grains. Presentation is variable in time and space, both among different species and within the same species (Laursen, 2001).

Release occurs when disseminules are detached (e.g. from flower anthers in the case of pollen) and entrained in wind flow. Fungi commonly use active mechanisms to release spores. Some angiosperms actively release or propel seeds, but these seeds are seldom entrained for long in wind flow because of their weight (Greene and Johnson, 1995). Passive release of disseminules operates according to the same mechanisms as for any particle (e.g. soil). Lift and drag caused by the wind interacting with the particle are resisted by the weight of the particle, cohesion between particles (or between, for example, a pollen grain and the anther), electrostatic forces, and the mechanical packing of particles with one another (see Section 8.2.1). Greene and Johnson (1995) have given a mechanistic explanation of seed abscission and suggested that abscission is probably proportional to drag, which itself is a function of the square of the wind speed. Because of boundary layer effects, typical mean wind speeds may not be sufficient to cause release of many disseminules. Aylor and Parlange (1975) and Aylor (1990) suggested that wind gusts are critical.

Once aloft, disseminules are subject to gravitational settling. This sedimentation is complicated by the irregular shape of disseminules, difficult-to-measure densities, variations in density with humidity, variability in fall orientation, and wake effects (Chatigny *et al.*, 1979). Furthermore, sedimentation rates in the air are offset or amplified by the vertical component of turbulent eddies. While sedimentation rates can be calculated using Stokes' law (Eq. 8.1), the above complications have led many modelers to adopt empirical measures (Jackson and Lyford, 1999). Despite modeling difficulties, it is clear that the total distance a disseminule is transported is partly dependent on its settling rate.

Jackson and Lyford (1999) reviewed pollen dispersal modeling in the context of Quaternary plant ecology. Because fossil pollen assemblages are used to reconstruct probable past distributions of vegetation, it is essential to understand the mechanisms by which these assemblages arise. Jackson and Lyford noted that most formal models of pollen dispersal applied in the paleoecological context (e.g. Tauber, 1965; Kabailiene, 1969; Prentice, 1988; Sugita, 1993) were based on Sutton's (1947) equations for diffusion in turbulent air, where pollen is released either at the ground level (Prentice, 1985) or from raised canopies (Kabailiene, 1969). The pollen plumes disperse downwind as a function of source strength, wind speed, turbulence, a vertical diffusion coefficient, the height of the emitting source, and the deposition velocity of the pollen particles (Jackson and Lyford, 1999). According to Jackson and Lyford, models to date have treated atmospheric parameters as constants, assuming neutral atmospheric conditions. They argued that more realistic treatment of atmospheric conditions for pollen transport is needed because pollen can disperse more widely in unstable atmospheric conditions.

Spatially explicit models of pollen dispersal include distributed pollen source areas and wind fields. Kawashima and Takahashi (1995) modeled the potential concentration of allergy-aggravating cedar pollen in Japan using this approach. Their model was based on a raster map of cedar density and hourly meteorological data, which included a generated wind field. A pollen emission submodel based on the relationship between pollen emission rates and meteorological conditions provided the pollen source strength to a pollen distribution submodel, which considered downwind transport and Gaussian diffusion. A vertically distributed model was used by De Jong *et al.* (1991) to examine the probability of undesired escape of fungal spores deployed in biocontrol of *Prunus serotina* in conifer forests. Their model used four vertical layers and calculated turbulent diffusion of spore fluxes in each layer and the horizontal and vertical "escape rate" of fungal spores from the model domain.

SIGNAL TRANSPORT IN WIND

Some plants, when attacked by herbivores, emit volatile chemicals that signal animal predators to rally to the defense of the plant by preying on the herbivores (Kessler and Baldwin, 2001). Presumably to attract mates, gypsy moths (and many other animals) emit "puffs" of pheromones into the air (Elkinton *et al.*, 1987). These and other odors transported by wind contain information when a receiving organism translates the odor signal into a stimulus that initiates some behavior (*i.e.*, flying towards a plant or a mate).

Odors are most often modeled as plumes originating from point sources, and their treatment is exactly analogous to (and usually based on) the methods used to simulate transport of airborne contaminants (see discussion above). Models are often one dimensional, in the sense that they predict the trajectory and spread of the odor-producing chemical as a function of distance from the source, without explicit regard for direction (downwind transport is assumed). Models typically consider (i) advective transport of the chemical imbedded in the wind flow; (ii) dispersion caused by the turbulent mixing of eddies (Elkinton and Carde, 1984; Elkinton *et al.*, 1984); and; (iii) molecular diffusion along concentration gradients (see Ch. 6). Elkinton *et al.* (1987), for example, used Gaussian plume models to predict the dispersion of gypsy moth pheromones. They contrasted the trajectory of a gypsy moth pheromone plume beneath a forest canopy with that over a grassy field (reported by David *et al.*, 1982) and noted that in nature the moths typically are found in forest understories.

We have touched here on only a handful of examples of wind-transport modeling in order to provide an introduction to some of the strategies that are used. A good starting place for readers interested in pursuing this further is the literature collection cited in the text. The demonstration model, Snow Tran-3D, described in the next section provides an opportunity for hands-on exploration of one type of wind transport.

8.4   Introduction to the example model

An example of aeolian transport is the redistribution of snow in arctic and alpine environments by wind interacting with topography and vegetation. The result is erosion of snow from areas experiencing high wind speeds, and deposition of snow where wind speed is reduced. The redistribution of snow across the landscape results in spatial differentiation of ecosystem properties, including radiation balance (Liston, 1999), chemical inputs (Williams et al., 1996b), meltwater (Billings and Bliss, 1959), rock weathering (Benedict, 1993), pedogenesis (Holtmeier and Broll, 1992), decomposition and mineralization (Williams et al., 1998), animal habitat (Coulson et al., 1995), and vegetation (Billings, 1969; Walker et al., 1993). Hiemstra and colleagues (Hiemstra, 1999; Hiemstra et al., 2002) simulated snow redistribution in the Medicine Bow Mountains of Wyoming, and much of the discussion here is derived from their work.

The importance of snow redistribution by wind in arctic and alpine environments cannot be overstated, but monitoring and measuring this redistribution is difficult. While remote sensing provides information about the spatial extent of snow (Baumgartner and Apfl, 1994; Walsh et al., 1994), it cannot directly measure snow depth. Field measurement of snow depth is possible (Goodison et al., 1981) but laborious (Fig. 8.10), and frequently requires sampling in harsh environments during winter (Hiemstra, 1999; Hiemstra et al., 2002). Another approach is to model snow depth and extent by combining empirical data and physical principles. To demonstrate modeling of wind as a transport vector, we describe an implementation of the Snow Tran-3D model (Liston and Sturm, 1998) for simulating snow redistribution in the Medicine Bow Mountains of Wyoming.

8.4.1 Spatial domain

The implementation of Snow Tran-3D provided on the CD with this book is parameterized and modified for simulation of snow transport over an area known as Libby Flats, in the Medicine Bow Mountains of Wyoming. Libby Flats is a broad, open ridge just above the treeline and ranges from 3100 to 3300 m in elevation (Hiemstra et al., 2002; Fig. 8.11). Vegetation includes alpine meadows and isolated patches of Engelman spruce and subalpine fir, some of which, in the most windy locations, have taken on a "krummholz" structure. Topography on Libby Flats is rolling, with low-relief hills and swales.

FIGURE 8.10
Chris Hiemstra measuring snow depths on Libby Flats in the Medicine Bow
Mountains, Wyoming. Field measurement of snow conditions is laborious and
often requires working in harsh climatic conditions. (Photograph by Theresa Kay.)

FIGURE 8.11
A view of Libby Flats in the Medicine Bow Mountains of Wyoming. This high,
windswept ridge at and above the treeline serves as an excellent test domain for the
Snow Tran-3D model. Notice the snow deposition alee of trees on Libby Flats.
(Photograph by Chris Hiemstra.)

To adapt Snow Tran-3D to Libby Flats, changes were made in model resolution, subgrid parameterization, and meteorological forcing data (Hiemstra
et al., 2002). Specifically, cell size was reduced from 20 m × 20 m to 5 m × 5m.
The vegetation present on Libby Flats was included in the spatial domain and
given one of two snow-holding capacities depending on height. Subgrid parameterization included addition of a smoothing function to prevent snow deposition downwind of objects from exceeding the height of the object, a condition
that does not occur in the real world. Also, meteorological data obtained from
other geographic locations were corrected to represent more accurately the conditions actually found on Libby Flats.

The model, written in FORTRAN, reads input data files, runs iteratively
for each day in the temporal domain, and produces an output file containing
a matrix of snow depths, one for each raster cell in the spatial domain. The
output matrix is converted to a grid image that is displayed and queried in
ArcView.

8.4.2   Principles of model

Snow Tran-3D uses physical principles to simulate snow redistribution temporally and spatially over heterogeneous landscapes (Liston and Sturm, 1998; Hiemstra *et al.*, 2002). Model inputs include non-spatial data describing solar radiation, precipitation rate, wind speed and direction, air temperature, humidity, and vegetation snow-holding capacity, along with a gridded (raster) representation of the spatial domain including topography (elevation) and the distribution and height of vegetation. Meteorological data are updated on a daily time step to allow calculation of daily snow distribution and depth.

To predict snow redistribution, Snow Tran-3D first calculates a wind field across the spatial domain using wind speed and direction and the spatially distributed data describing elevation and vegetation. Topographic obstructions and vegetation alter the wind field as the simulated wind moves over and around them. Surface shear stress is determined from the wind field; above a certain threshold, shear stress dislodges snow particles to initiate transport (Kind, 1981; Sturm *et al.*, 2001). Availability of snow also plays an obvious role, and the amount of snow must be greater than the vegetation snow-holding capacity for transport to occur.

In the model, snow moves by saltation and/or turbulent suspension of snow particles; snow is added or removed by snowfall, transport, and sublimation (Hiemstra *et al.*, 2002). The change in snow quantity for individual grid cells must satisfy a mass-balance equation, with saltation, turbulent suspension, sublimation, and precipitation balancing one another (Liston and Sturm, 1998; Liston, 1999). Sublimation occurs only during transport. Finally, when the wind field is strong enough to cause transport within the model domain, it is assumed that snow enters the domain by transport from upwind areas outside of the domain and leaves on its downwind edge.

8.4.3   Model limitations and deficiencies

Snow Tran-3D is a complex model that necessarily requires simplifying assumptions at many levels. According to Liston and Sturm (1998):

> The model performs reasonably well in our test cases, but it is still a simple approximation to a complex natural system. Model deficiencies can be divided into two areas: those that arise because of limitations and constraints on input data, and those that arise due to erroneous assumptions, over-simplifications, or errors in the model snow-transport physics. The former limitations are thought to be more severe than the latter.

Input data limitations include the resolution of the topographic data (inability to resolve topographical features smaller than grid units) and the spatial and temporal resolution of meteorological data (failure to resolve fine-scale or short-lived weather anomalies). These limitations could be significant; for example, when subgrid topographic features present in the landscape but not depicted in the input data interact with the wind to affect snow depths. Spatial resolution also interacts with model physics when the shapes of topographic features change with resolution. Similarly, because meteorological data are averaged over time, the importance of intense but brief periods of high wind may not be reflected in model results.

Liston and Sturm (1998) also discussed simplifications in the physical formulations of Snow Tran-3D that affect model results. Snow density and the shear-velocity threshold for dislodging snow are constant across the spatial domain of the model. This has obvious effects on snow redistribution if spatial and temporal heterogeneity in these properties is important. Sensitivity tests performed by Liston and Sturm in the Arctic suggested that this simplification did not cause substantial differences in model output. In contrast, alpine systems at lower latitudes experience variation in solar radiation that can substantially affect snow density on different aspects.

The model also does not simulate feedbacks between sublimation and humidity. One expects that humidity will increase as snow sublimates; in turn, this increase in humidity will slow the rate of sublimation. In the model, no such feedback occurs (Liston and Sturm, 1998). The model authors suggest that the magnitude and direction of error introduced by this simplification is unknown. Finally, Liston and Sturm (1998) stated that Snow Tran-3D assumes a "vertical distribution of mean particle size," when in fact "particle sizes within the snow-transport profile can vary by an order of magnitude." They further state that the model treatment of snow particle size "is expected to influence many of the processes associated with turbulent kinetic energy and moisture transfers occurring during snow transport."

Hiemstra et al. (2002) modified Snow Tran-3D for a near-treeline alpine landscape in Wyoming and gathered snow-depth data in the field for validation (see below Figs. 8.10 and 8.11). Hiemstra found that the model overpredicted snow depths in this area where snow was shallow, and underpredicted them where it was deep. Hiemstra (1999) attributed these model errors to scaling considerations, difficulties in simulating the effects of trees, spatial errors, snow conditions, and the representation of meteorological data. Liston and Sturm (1998), in their Arctic work, found that the model was able to describe general patterns but diverged with respect to some details of the snow distribution and depth.

8.5   Summary

Transport by wind is important in almost all ecosystems and occurs across a wide range of spatial and temporal scales. The physics and mechanisms of atmospheric transport have been studied extensively, perhaps because of their relevance to human well-being. Models include characterizations of fine-scale phenomena, like the entrainment of individual particles into the wind stream, and global systems, like coupled ocean–land surface–atmospheric transport of energy within and between hemispheres. Despite the sophistication and complexity of these models, most of them are zero dimensional, some are one dimensional and only a few are two dimensional. Much work remains to be done in explicitly incorporating spatial heterogeneity.

9

Fire

The worst thing that we could have is to be so enamored of our forests that we eliminate the processes that change them.

Tom Atzet, US Forest Service ecologist for southwestern Oregon
(quoted by Jeff Barnard, Associated Press (2002) in an article on
the Biscuit Fire in Oregon)

9.1   Transport system description

Norman Maclean's book *Young Men and Fire* (1992), published after his death in 1990, is a testament to the awe humans feel for the power of fire and to our fascination with the processes that create and sustain it. The book describes physical, ecological, and human aspects of the Mann Gulch fire in Montana, which, in the summer of 1949, killed 13 fire fighters caught in a "blowup." As a tribute to the men who died in Mann Gulch, Maclean elegantly described the processes leading to the conflagration. A passage from the book (p. 35) captured the sense of fire as an irresistible force on the landscape:

> The crown fire is the one that sounds like a train coming too fast around a curve and may get so high-keyed the crew cannot understand what their foreman is trying to do to save them. Sometimes, when the timber thins out, it sounds as if the train were clicking across a bridge, sometimes it hits an open clearing and becomes hushed as if going through a tunnel, but when the burning cones swirl through the air and fall on the other side of the clearing, starting spot fires there, the new fire sounds as if it were the train coming out of the tunnel, belching black, unburned smoke. The unburned smoke boils up until it reaches oxygen, then bursts into gigantic flames on top of its cloud of smoke in the sky. The new firefighter, seeing black smoke rise from the ground and then at the top of the sky turn into flames, thinks natural law has been reversed. The flames should come first and the smoke from them. The new firefighter doesn't know how his fire got way up there. He is frightened and should be.

Perhaps the enduring human captivation with fire speaks to its complexity more than any detailed physical description; however, fire's enormous human and ecological impact has encouraged substantial research into fire physics, fire propagation mechanisms, and strategies for modeling fire predictively and in "real time." Integration of physical processes across scales, from the ignition of fuel to the influence of meteorological factors, is necessary to simulate fire behavior effectively. In this chapter, we discuss fire as a transport system. Initiating events and conditions, transport mechanisms, and strategies for integrating these to model fire behavior across landscapes provide another perspective for thinking about transport.

9.2   Fire as a transport phenomenon

In a sense, fire is different from other transport vectors because the transported entities arise from the vector itself. Like the crest of a flood propagating downstream, fire burning across space leaves a path of influence behind it, rather than at its point of termination. Fire is a self-sustaining reaction that moves wave-like across land, converting fuels and oxygen into energy, solid residues (e.g. ash), and gases (e.g. smoke, water vapor), which themselves can be transported by other vectors (e.g. wind and water). Fire is a special case of wind transport to the extent that it is sustained by wind and the path it follows is partly directed by wind. However, the unique physical nature of fire, and its profound ecological effects (see Section 9.3.4), cannot be attributed to wind alone. For these reasons we treat fire here as a unique transport vector, capable of moving matter, energy, and information across space, and with important consequences.

The energy released when fuels are burned by fire is the most obvious and, arguably, the most important entity transported by fire. It is this energy that causes most of the ecological, geomorphological and atmospheric effects that come to mind when we think of fire. Other entities transported by fire result, directly or indirectly, from the energy released during combustion. Clarke *et al.* (1994) explicitly recognize primary and secondary transport of materials by fire, noting that "fires are a major source of gases and particulate in the atmosphere, including hydrocarbons, carbon dioxide, and ammonia." All can affect climate (Andreae, 1991). Secondary transport occurs when fire sets the stage for transport by other vectors. Examples are erosion of burned areas resulting in transport of nutrients and sediment into lakes and streams (Clarke *et al.*, 1994; Ch. 10), changes in patterns of animal movement (Ch. 11), and weakening of slopes, leading to slope failure (Benda and Dunne, 1997; Ch. 7).

9.3   Underlying principles for mode of transport

9.3.1 Initiating events

Fire initiation requires the presence of fuel and an external source of activation energy sufficient to cause pyrolysis (Whelan, 1995; Pyne *et al.*, 1996). Fire begins when an event, like a lightning strike, occurs at a place where chronic conditions, like fuel accumulation and drought, have created the circumstances necessary for combustion. To understand initiation of transport by fire, one must understand the spatial distribution and physical nature of initiating events, weather, and fuel, plus the physics and chemistry of combustion, and, importantly, the interaction of these components (Whelan, 1995).

Fire has been part of the terrestrial environment since prehistoric time, mostly resulting from ignition by lightning and volcanism, but also caused by humans, who have altered its frequency and distribution (Pyne, *et al.*, 1996). The use of fire for hunting and clearing undergrowth in historical times is well documented (Gruell, 1985; Knight, 1994; Whelan, 1995). Russel (1983) reviewed documents from the sixteenth and seventeenth centuries that included first-hand descriptions of widespread and frequent use of fires by Native Americans for agricultural clearing, hostile actions, signaling, hunting, and increasing grass productivity. Prescribed burns are still used as management tools (e.g. Kolata, 1984) and occasionally wildfires are started when prescribed burns get out of control.

Globally, humans are the most common ignition source for fire (Whelan, 1995; Pyne *et al.*, 1996; Vasconcelos *et al.*, 2001). Arson, prescribed burns, and accidents are common anthropogenic initiating events. Lightning is the most widespread non-anthropogenic ignition source (Whelan, 1995) and dominates fire initiation in some parts of the world, including the western USA, where two-thirds of fires are started by lightning (Pyne *et al.*, 1996). Other less-common natural sources include volcanic eruptions and sparks generated during rock slides.

It is estimated that, on average, the earth receives 100 lightning strikes per second (Pyne *et al.*, 1996), with some of these in potentially flammable places. During a single storm in southern Oregon and northern California in 1987, 1600 fires were ignited by lightning (Pyne *et al.*, 1996). Lightning simultaneously affects the properties of fuel and supplies activation energy. When lightning strikes a tree, for example, a shower of bark, wood, and needle particles is produced along with a cloud of volatile extract (Fig. 9.1). These finely divided fuels are almost explosively flammable; if fuel conditions on the forest floor and in the canopy are suitable, the fireball caused by their ignition can ignite larger fuels and initiate combustion.

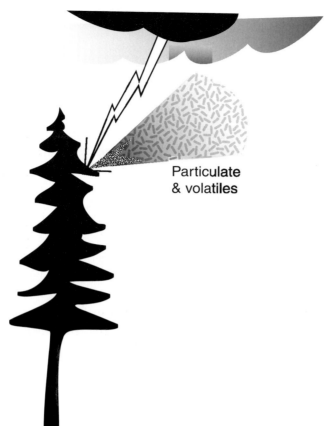

Particulate
& volatiles

FIGURE 9.1
Lightning strikes are a
common cause of ignition
leading to wildfire. When
lightning strikes a tree, an
almost explosive mixture
of fine fuels and volatilized
flammable chemicals are
simultaneously released
and ignited. If nearby
fuel conditions are suitable,
these too will ignite and
initiate wildfire.

Fire itself initiates combustion both at the flame front (see below) and via
burning debris transported by wind through the air to locations some distance
from the extant fire (Fig. 9.2). Whelan (1995) reported studies (Foster, 1976;
Vines, 1981) documenting initiation of fires by windborne firebrands 25–30
miles from the main fire. Fire is self-sustaining as long as flammable fuels are
available at the fire front or in the downwind path of firebrands.

Fuel conditions

Fire initiation (and propagation) depends on the fuel properties at the place
where activation energy is supplied (e.g. by lightning). The factors determining the spatial distribution and physical and chemical properties of fuels are
as complex as those determining these properties of plants and vegetation.
Vegetation, disturbance history, and decomposition rates are but some of the
relevant ecosystem properties (Mooney *et al.*, 1981; Pyne *et al.*, 1996). Horizontal
and vertical heterogeneity in fuel accumulation affects the spatial distribution

FIGURE 9.2
The complex pattern of burned and unburned lodgepole pine forest in Yellowstone
National Park after the 1988 fires. The small burned area among unburned trees in
the foreground is probably the result of "spotting," where airborne embers ignite
fires that are isolated from the main fire. (Photograph courtesy of Monica Turner.)

of the probability of fire initiation and the burn pattern when fires do start
(Pyne *et al.*, 1996). This, in turn, affects the fuel accumulation pattern after a
fire (Fig. 9.3). These complex feedbacks are important structuring processes in
many ecosystems (Knight, 1994).

Rundel (1981) and Pyne *et al.* (1996) discussed fuel flammability with regard
to structural and chemical properties of the fuel materials. They elaborated
on the importance of structure, including fuel loading, particle density, fuel
surface area to volume ratios, and fuel porosity and they discussed the relationships between the chemical composition of fuels and their flammability. Rundel emphasized that fuel moisture content is of critical importance.
Water in fuel raises the activation energy needed to initiate fire because heat
is required to raise the temperature of fuel water to boiling, separate bound
water from the fuel, vaporize the water, and heat the water vapor to flame temperature (Whelan, 1995). Meteorological conditions, particularly drought and
heat, reduce fuel moisture and increase the probability that an initiating event
will supply the needed activation energy for ignition.

FIGURE 9.3
The distribution of fuels can be heterogeneous in both vertical and horizontal dimensions. The dead branches in the foreground provide a "fuel ladder" that can carry flames up into the crown of a burning forest. This patch of dead trees represents an isolated concentration of fuel in the horizontal dimension.

9.3.2 Fire spread and intensity

The firefighters killed in Mann Gulch were overtaken by flames exploding up a steep hillside faster than they could run (Maclean, 1992). Fire spread can be terrifyingly rapid or tediously slow (Table 9.1). Pyne (1982) has given examples of fires (e.g. the Matilja fire of 1932 in Southern California) that burned tens of thousands of acres in just a few hours, advancing at up to 5 mph (8 km/h). The 1988 fires in Yellowstone National Park reportedly advanced at rates of 2 mph (3.2 km/h), and at times faster. The factors determining the rate and spatial pattern of fire spread and the intensity with which fires burn are at the core of fire behavior modeling. These factors include characteristics of the fuels (discussed above), environmental conditions, and their spatial and temporal juxtaposition (Pyne *et al.*, 1996) (Fig. 9.4).

Pyne (1982), in his book emphasizing the cultural aspects of fire, contrasted "free-burning" fire, like that occurring in the wild, with the "controlled" fire

TABLE 9.1. *Examples of spread rates for fires in various fuel and wind situations*

Rate of spread (m/s)	Fire conditions	Comparison with human travel
5.1×10^{-3}	Litter fire with no wind on level ground	One person building fire line with heavy fuel
1.2×10^{-1}	Aged medium slash on 45° slope	Backpacker hiking up 45° slope
1.27	Low sagebrush with moderate wind	Brisk walk on level ground
4.08	Chaparral with moderate wind	Slow to moderate running pace
6.12	Dry grass with high wind	Runner doing 4 minute miles

From Pyne (1995; Fig 2.14).

FIGURE 9.4
A burn in Montana showing the effect of topography and wind direction on burn patterns. This fire burned up one aspect of the obvious ridge but was unable to "turn the topographic corner" to burn other aspects. Had wind been blowing from left to right in the photograph, the rightmost aspect may have been ignited.

harnessed by engineers in furnaces or engines for use by humans:

> Free-burning fire, however, is vastly more complex, its physics more statistical, and, in contrast to the fires of the hearth and furnace, its engineering efficiency much reduced. The parameters of the fire environment cannot be finely tuned to sustain maximum efficiency or

to modulate the wild fluctuations of energy release and mass transfer. Instead of a single fuel element, wildfire responds to an ensemble of fuel complexes, grossly arranged; instead of a furnace designed to ensure maximum draft, it must follow broad topographic configurations; instead of a metered intake of oxygen, it must deal with traveling air masses superimposed over microclimates; and instead of careful engineering to ensure maximum heat transfer, wildland fire is propagated by a variety of mechanisms, often erratic and turbulent.

The environmental heterogeneity encountered by "free-burning" fire, and described so well by Pyne, is central to understanding and modeling fire spread. Fire initiation is usually from single or multiple point sources; in the latter, the pattern and proximity of the sources determine whether the resulting fires act independently or synergistically (Pyne, 1982). Once a fire is burning, spread and intensity are influenced by available fuel, fuel moisture and temperature, chemical properties of the fuel, fuel continuity, wind, and topography (Whelan, 1995). We concentrate our discussion here on factors contributing to the spatial heterogeneity of fire spread.

Weather
Weather influences fire at every stage and across scales (Pyne *et al.*, 1996). Fire regimes shift in response to changes in global climate. Vegetation growth associated with an unusually wet season one year leads to large fuel volumes the next. Drought and heat during the fire season predispose fuels for ignition. Lightning, produced by convective storms, provides ignition. Wind, temperature, humidity, and precipitation affect the behavior of fires once they are burning. Finally, precipitation, humidity, and cooling temperatures slow and extinguish fires by increasing the energy required to sustain combustion. A discussion of fire weather across all of these scales would require a chapter of its own, and we defer here to the excellent treatment by Pyne *et al.* (1996). We do, however, discuss wind in more detail (see below) because of its strong influence on fire behavior, both physically and spatially.

Wind
Fire creates its own wind at local scales and is strongly affected by environmental (ambient) wind across scales. In this sense, transport by fire may be thought of as a special case of wind transport, although the unique physical properties and ecological effects of fire warrant special treatment. As fire burns, it creates convection: heated air rises above the flame. Surrounding air is drawn inward to replace the convected air, creating an "indraft" and supplying oxygen to the fire (Whelan, 1995; Pyne *et al.*, 1996). Fire spreads outward as fuel in the vicinity

FIGURE 9.5
Fire creates its own wind as oxygen in the immediate vicinity of the fire is burned during combustion and as heated air moves upward. Air at the margins of the flame moves toward the flame, thus creating wind. Fire tends to burn preferentially in a downwind direction because flames bent over by the wind preheat fuels downwind more quickly, leading to more rapid downwind ignition. Similarly, fire burns more quickly up slopes because the flame is closer to the upslope side, and fuels there are preferentially preheated. (Modified from Whelan, 1995.)

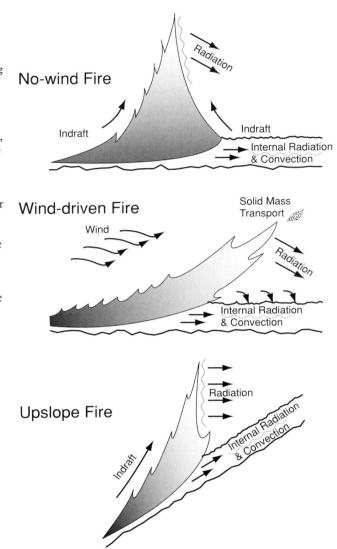

of the existing fire is preheated and eventually ignites. The self-created wind provides a constant source of oxygen to support ignition of new fuels.

Ambient wind affects the rate of fire spread by several mechanisms. First, wind pushes the flame over, reducing the angle between it and the ground (Fig. 9.5) and increasing its ability to preheat fuel downwind (Whelan, 1995; Pyne *et al.*, 1996). Second, the wind can carry burning fragments of fuel, creating "spot" fires when the fragments ignite fuel in new areas. As discussed earlier, spot fires can arise long (or short) distances from the main fire and burn independently or synergistically. Third, ambient wind supplies oxygen to the fire and increases its ability to burn fuel, much as blowing on a campfire enhances the flames. Finally, ambient wind can intensify the removal of moisture from fuels over time, increasing the likelihood that ignition will occur.

Changes in wind direction during a fire have strong effects on the burn pattern. Whelan (1995) described a grassland fire in Nebraska that burned eastward for 18 km along a 5 km swath before the wind changed and caused the fire to burn northward along the 18 km edge. In this case, the lateral edge of the fire became the fire front because of the wind change. Changes in local winds are important; they can arise from interactions of broader-scale meteorology with topography and vegetation. None of the factors that influence speed and direction of fire spread operates in isolation.

Topography

Topography and its interaction with other environmental features, like wind, vegetation, and fuel conditions, affect the burn pattern, spread rate, and intensity of fires across a landscape. All other factors being equal, fire tends to burn more rapidly upslope for the same reasons that it advances faster downwind (Pyne *et al.*, 1996) (Fig. 9.5). As a fire burns upslope, the flames are closer to the ground on the uphill side, allowing preheating of fuel to occur more rapidly there than on the downhill side. The intense thermal radiation from the fire front preheats and dries the upslope fuels. The Mann Gulch fire fighters suffered when the fire rushed up the slope from below, overtaking them. Topography also creates natural fire breaks, slowing or stopping fire spread. Bare ridgetops, rocky cliffs, rivers, and lakes are all topographic features that serve as boundaries for fires. Of course, fire can "jump" these boundaries if they are narrow and airborne flaming debris blows over them, igniting new fuel across the boundary.

Topography can also have indirect effects on fire. In the northern hemisphere, north-facing slopes receive less sunlight than south-facing slopes and, as a consequence, are often cooler and moister. In mountainous areas, topography and snow interact with wind to create snowdrifts downwind of topographic rises (Ch. 8). As the snow melts from these accumulation areas, it creates pockets of relatively moist, fire-resistant habitats (Billings, 1969). Finally, as discussed for wind (Ch. 8), topography alters the structure of the wind field, which itself directly affects fire behavior.

Fuel distribution

It is clear that understanding the horizontal and vertical distribution of fuel is central to understanding the spatial pattern of fire spread and intensity (Pyne *et al.*, 1996). Yet this distribution is as complex as terrestrial vegetation ecology, since the very factors that determine vegetation distributions (including fire history) ultimately result in the production of fuels. Further, horizontal and vertical distribution are scale dependent. Generalized fuel properties are associated with scales of vegetation pattern. But even within vegetation types, horizontal and vertical distributions vary with unique stand properties, such as the presence or absence of snags and downed logs, or very local events, such as individual canopy falls. We touch on fuel properties above (p. 166) and refer interested readers to Rundel (1981) and Pyne *et al.* (1996) for more detailed discussions.

9.3.3 Fire extinction

When a fire has burned all available fuel or can no longer supply the activation energy necessary to ignite new fuels, it is extinguished. Extinction results from decreased fire intensity or changes in fuel conditions. Precipitation, humidity, cooling ambient temperatures, and decreased wind act alone or in concert to increase the chances that a fire will die. All of these conditions are encountered in different places in the affected landscape at different times, thus altering the configuration of the burn.

Natural or human created obstructions can prevent a fire from encountering new fuel. Firefighters create conditions that discourage fire spread by wetting fuels, thus raising the activation energy needed for ignition to occur, and by building artificial fire breaks. Natural conditions (e.g. topography; see above) can also create fire barriers. Active fire suppression by humans has been practiced for over a century in the USA (Pyne, 1995; Pyne *et al.*, 1996) and longer in other parts of the world, often resulting in unnaturally high fuel availability and the potential for larger and more intense fires than might be expected under natural fire regimes (Baker, 1992; Knight, 1994).

9.3.4 Ecological effects of fire

Fire has important effects on the ecology and geomorphology of the areas over which it burns (Pyne *et al.*, 1996). However, at broader scales, environments are affected by gases and particulate matter emitted by fire (Crutzen and Andreae, 1990). Smoke degrades air quality locally and regionally, and gaseous emissions (e.g. CO_2) affect climate globally by contributing greenhouse gases (Levine, 1991). In most terrestrial ecosystems (arctic tundra and tropical cloud forests are notable exceptions), fire has a significant influence on animals (Bendell, 1974; Singer *et al.*, 1989; Pyne *et al.*, 1996), vegetation (Barbour *et al.*, 1987; D'Antonio and Vitousek, 1992; Knight, 1994; Archibold, 1995), biogeochemistry (Schlesinger, 1997; Crutzen and Andreae, 1990), and the physical environment, including soil erodability (Hungerford *et al.*, 1991) and slope stability (Benda and Dunne, 1997).

In many ecosystems, including boreal forests, montane conifer forests, Mediterranean shrublands, tropical savannas and temperate and tropical grasslands, fire plays an intrinsic role in ecosystem structure and function (Barbour *et al.*, 1987; Collins, 1992; Johnson, 1992; Archibold, 1995). In areas where fire is frequent, some plants have evolved to depend on it for reproduction (Knight, 1994; Romme *et al.*, 1997). Lodgepole pine (*Pinus contorta*), for example, common in the northern Rocky Mountains of the USA and southern Canada, produces serotinous cones that require the intense heat of fire to open and release seeds (Knight, 1994). This strategy increases the chances of seeds finding suitable conditions for lodgepole establishment: mineral soil and low competition for light and nutrients (Knight, 1994).

The structure of landscapes can be controlled by the spatial and temporal variability of periodic fire (Romme, 1982; Turner *et al.*, 1994a; Pyne *et al.*, 1996). Because the factors contributing to combustion (fuel availability and composition, topography, wind, etc.) vary across space, fire burns with variable intensity at different places on the landscape (Fig. 9.2). A burned forest, for example, may include areas where fire exploded into the forest canopy killing all vegetation; adjacent to these areas there may be areas where fire raced through understory growth at low intensity, leaving trees essentially unharmed (Knight, 1994). Furthermore, areas that have burned recently are less likely to burn again than unburned areas. This results in patches of vegetation at different successional stages (Romme, 1982; Dale *et al.*, 1984; Turner *et al.*, 1994a).

In ecology (and the popular press), fire is more often treated as an ecological disturbance than as a transport vector (e.g. Pickett and White, 1985; Dale *et al.*, 1986). In a sense though, "disturbance" is a catch-all, encompassing many disparate *effects* of the transport of energy, matter, and information by the fire vector. For example, the resetting of succession by fire includes changes in species richness, biotic diversity, productivity, and competitive relationships, to name a few (West *et al.*, 1981). Other effects grouped into the term "disturbance" include alteration of animal habitat (Singer *et al.*, 1989; Turner *et al.*, 1994b), soil properties (chemical, biological, and physical) (Halvorson *et al.*, 1997), erosion susceptibility (Hungerford *et al.*, 1991), grazing patterns (Noy-Meir, 1995), insect outbreaks (Gara *et al.*, 1985; Anderson *et al.*, 1987), and plant and animal reproductive strategies (Knight, 1994; Romme *et al.*, 1997). All of these ultimately result from the transport of entities, especially energy, by fire.

9.4 Modeling techniques

Fire modeling is complicated by the interaction of many environmental factors that are themselves variable in time and space and, therefore, difficult to anticipate. Pyne (1982; p. 21) described the complexity of fire modeling in the following passage:

> Wildland fire behavior multiplies probability with probability. Unlike astronomy, where it is possible to predict the position and velocity of individual objects with great precision, fire behavior deals with statistical ensembles – the limitless nuances of fuel complexes, the restless variety of topographic forms, and the maddening vagaries of weather, particularly on micro- and mesoscales. Wildland fire does not merely resemble a climatic or meteorological phenomenon; it results from them and is thus another order removed from simple determinism. Rather than presetting the key parameters, it is necessary to predict what they will be.

Fire modeling is typically concerned with predicting the rate and direction of fire spread (e.g. Finney, 1993), the intensity with which a fire will burn (e.g. Rothermel, 1972), the shape of the area that will ultimately be burned by a fire (e.g. Richards, 1993, 1995), and the likelihood of fire extinction (e.g. Latham and Rothermel, 1993). Some models proactively estimate fire hazard based on fuel loads, topography, and wind patterns in relationship to the existence of human habitation (e.g. Deeming *et al.*, 1977; Chuvieco and Congalton, 1989). Still others characterize the fire regime of particular places (e.g. Davis and Burrows, 1994). We concentrate our discussion here on models that simulate fire spread and intensity, including the processes of ignition and extinction.

9.4.1 Types of fire model

Davis and Burrows (1994) divided fire models into two categories: those based directly on the physics of fire behavior and those combining physical factors with stochastic processes. They also discussed fire models that use GIS and remotely sensed data. It is important, for our purposes, to distinguish between models that predict the probability of fire at a particular point (e.g. Andrews, 1986) and those that are spatially distributed (e.g. Clarke *et al.*, 1994; Hargrove *et al.*, 2000). Among the latter, Finney (1996) makes a fundamental distinction between non-cellular and cellular (cellular automata) models. Non-cellular models use continuous mathematical equations to simulate spread (e.g. FARSITE™ described below). Cellular models are based on raster grids with fire spreading from cell to cell. They are further categorized by the algorithm used for fire propagation (Finney, 1996), including deterministic spread, probabilistic spread, and fractal modification of spread rate and geometry. Non-deterministic models require multiple runs (a Monte Carlo approach) to generate "event-probability maps" for given scenarios.

Gardner *et al.* (1999) and Hargrove *et al.* (2000) reviewed fire modeling strategies with an emphasis on scale. They argued that many fire models are designed to predict local ground fires for limited time periods (hours) and are less suited for broad-scale simulations. They also suggest that cellular models based on probabilistic spread are "the most efficient" approach and are, therefore, well suited for broad-scale simulations. They mention, however, that most models using this approach consider only homogeneous scenarios of wind and fuel. Finney (1996) also emphasized this, noting that "none of the cellular models adequately simulated fire spread under test conditions with spatial and temporal heterogeneities." Recent cellular models (e.g. Clarke *et al.*, 1994; Hargrove *et al.*, 2000) may overcome some of these difficulties.

9.4.2 Non-cellular models

Physical models of fire behavior are usually based on equations presented by Rothermel (1972). These equations predict fire reaction intensity and spread rate as functions of fuel properties and meteorological factors measurable in the field. Meteorological factors include air temperature, humidity, wind speed, and wind direction. Fuel parameters include the amount of fuel per unit area, fuel size distribution (diameter), height of the fuel bed, fuel heat content, particle density, and fuel moisture content. A premise of Rothermel's model is that fire spread is a related series of fuel ignition events.

In the Rothermel model, fire spread rate and intensity are calculated for "a continuous stratum of fuel that is contiguous to the ground" without explicit consideration of spatial heterogeneity in fuel supply, vertical propagation of fire from the ground into forest crowns, or the effect of firebrands in allowing fire to leap ahead (Rothermel, 1972). Some of these limitations have subsequently been addressed. Van Wagner (1977) contributed to understanding crown fire behavior and Albini (1979) studied spotting caused by firebrands. The widely used BEHAVE model (Andrews, 1986; Andrews and Chase, 1989) relied on Rothermel's equations to predict the rate of spread, intensity, and flame length of fires for single points in time and space, but output was not spatially explicit. This model can estimate the total burn area if the time since ignition is known. A recent update of BEHAVE, called BehavePlus, adds routines to account for crown fire and improves the user interface.

Rothermel's equations, despite their limitations, have been adapted for models that consider spatial heterogeneity (e.g. Frandsen and Andrews, 1979; Vasconcelos and Guertin, 1992). Rothermel (1972) touched upon the spatial pattern of simple fires in continuous fuels and recognized that simplified scenarios are starting points for modeling fire spread in heterogeneous environments.

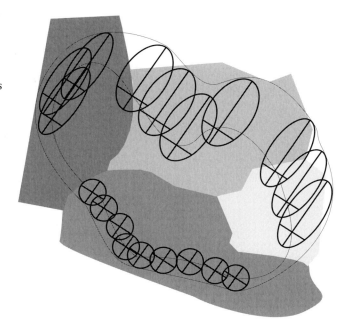

FIGURE 9.6
The elliptical burn scheme used in models like FARSITE™. The long axis of each ellipse is typically oriented parallel to the prevailing wind or up slopes. The size of the ellipses is influenced by these factors as well as by fuel conditions, indicated in the drawing by different shades of gray. More favorable fuel conditions lead to larger ellipses.

Most fires begin from a single source and spread outward, growing in size and assuming an elliptical shape with the major axis in the direction most favorable to spread. When the fire is large enough so that the spread of any portion is independent of the influences caused by the opposite side, it can be assumed to have stabilized into a line fire. A line fire behaves like a reaction wave with progress that is steady over time in uniform fuels.

Finney (1993, 1996) developed this in his "Fire Area Simulator" (FARSITE™) model, which predicted the spread and behavior of fires in heterogeneous terrain, fuel, and weather. The model was based on Huygen's principle (Anderson *et al.*, 1982; Richards, 1990) of wave propagation of the fire front, which is treated as "independent sources of small elliptical wavelets" forming a sort of envelope around the original fire front (Finney, 1996). The shape and orientation of each ellipse is determined by wind and slope, and size is determined by fuel conditions (Fig. 9.6). Richards (1995) expanded on the use of non-elliptical shapes to represent fire spread. Environmental data inputs for FARSITE™ are raster layers in a GIS, implying that the input data are cell based even though the fire spread model is not. In fact, any non-cellular model of fire spread in heterogeneous environments must include some means of depicting environmental heterogeneity, and raster depictions are typical.

The FIRE BioGeoChemical succession model (FIRE-BGC) (Keane *et al.*, 1996, 1997) is an ecological process model that combines disturbance models with models of stand development. FIRE-BGC uses a simplified version of FARSITE™ to create spatial depictions of fire intensity and flame length. These fire disturbance characteristics are subsequently used as input to ecological models that explore the interaction of disturbance in coniferous forest landscapes with processes such as tree regeneration, growth, and mortality.

9.4.3 Cellular models

Another approach to simulating the spread and intensity of fires is to divide the spatial domain into a grid and allow the fire to move from cell to cell using various algorithms (e.g. Stauffer and Aharony, 1992). Models employing this cellular automaton approach typically use Rothermel's equations to simulate burning *within* cells but use different strategies to allow fire to move *between* cells. According to Finney (1996), cellular models calculate the ignition times of regularly spaced cells and determine the position of the fire front by inference from cells that ignite simultaneously. Finney suggested that modeling problems arise as a result of distortion of the fire shape that is caused by forcing the fire into a gridded geometry. This, however, is really an issue of spatial resolution, and a fine-scale grid should ameliorate the problem. Hargrove *et al.* (2000) argued that, for broad-scale modeling (e.g. fires that burn for many square kilometers), cellular models are efficient. Furthermore, cellular models offer a convenient way to depict spatial heterogeneity by allowing each cell to have unique characteristics.

Probabilistic cellular models simulate fire spread by assigning probabilities of ignition to cells based on their fuel characteristics or other environmental properties (Vasconcelos and Guertin, 1992; Davis and Burrows, 1994; Hargrove *et al.*, 2000). Turner *et al.* (1989) modeled the spread of disturbance as a percolation process on a simple binary grid, with cells assigned to categories of either "flammable" or "non-flammable." The "Regional Fire Regime Simulation" model (REFIRES; Davis and Burrows, 1994) used this approach to explore the fire regime in California chaparral. In this model, fires are initiated by "igniting" some number of randomly chosen hexagonal cells from the domain. The date, time, and weather conditions when the ignition occurs are also randomly chosen. Once the fire has started, it burns within cells according to Rothermel's equations and spreads among cells using a contagious diffusion algorithm and spread probabilities based on fuel conditions, wind, and topography (Davis and Burrows, 1994). REFIRES does not allow fire spread by spotting (airborne firebrands).

The EMBYR model (Ecological Model for Burning the Yellowstone Region) uses 50 m×50 m grid cells to depict spatial heterogeneity in fuel type, fuel moisture, wind speed, and wind direction (Hargrove *et al.*, 2000). A diffusion algorithm (bond percolation) allows the fire to move from a burning cell to one of the eight cells in its immediate neighborhood as an independent event with a given probability. Hargrove *et al.* (2000) noted that below a critical probability threshold, fire is unlikely to propagate across the landscape. Probabilities are weighted depending on wind speed and direction, and distant cells can be ignited by airborne firebrands. At the conclusion of each model time step, burned cells are extinguished.

Clarke *et al.* (1994) argued that the double-ellipse model of Anderson (1983), used by many modelers (e.g. Finney, 1993), is only accurate under very restrictive conditions and that burn patterns in wildfires under natural conditions resemble patterns that are essentially fractal. They suggested that "three factors alone, the variation in the slopes and aspects of terrain, the non-uniform nature of the fuels and their moisture content, and the varying and turbulent nature of wind, would alter a perfect double ellipse beyond recognition in seconds." Their cellular automaton model implements these ideas and is the example model for this chapter (Fig. 9.7). The model uses an algorithm called diffusion-limited aggregation to simulate spread and is described in more detail below.

9.5   Introduction to the example model

The example model for this chapter, developed and described by Clarke *et al.* (1994) and refined by Clarke and Olsen (1996), is a cellular automaton (raster) model that uses information on fuel loadings, topography, and other environmental factors to predict spatial and temporal fire behavior. Remotely sensed data can be linked to the working model to enable fire managers to use it in near real time (Clarke *et al.*, 1994), although the example model on the CD does not allow this linkage. The model was initially calibrated using remotely sensed data gathered for a fire that burned in the San Dimas (California) experimental forest in 1986. It is written in the C programming language and compiled for UNIX workstations. For the CD with this book, we modified the model to operate in a Windows (PC) environment.

9.5.1 Principles of the model

The model is based on the notion that the spatial spread of fire resembles a process called diffusion-limited aggregation and results in "objects" (the fire and the area affected by it) that have characteristics of fractals (Clarke

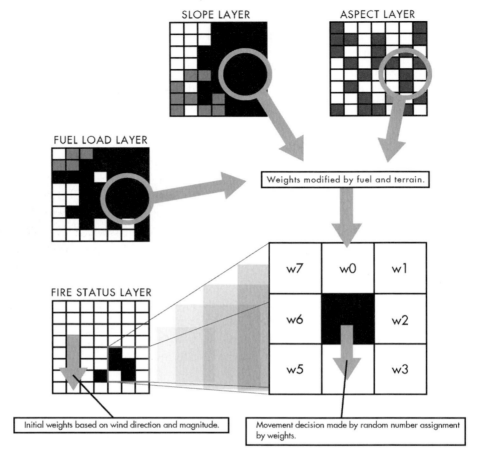

FIGURE 9.7
Schematic of a cell-based fire model showing the factors that influence the movement of "firelets." (Modified from Clarke *et al.*, 1994.)

et al., 1994). According to this scheme, objects are influenced by many random events, which are themselves influenced by preceding events. The model simulates fire spread using "firelets" originating from areas already burning. The firelets move based on consideration of other data layers (e.g. topography, wind patterns) that provide weights for movement rules (Fig. 9.7). Once a firelet has moved, it can survive if new fuel is ignited.

The model user initiates fire at any cell in a raster grid representing the model domain. Once started, the fire moves to an adjacent cell whose location is chosen using a random number weighted by the environmental layers described above. This process continues and the fire "runs" from cell to cell until the edge of the domain is reached, a distance threshold is surpassed, or no additional fuel is encountered. If successive fire runs do not encounter new fuel, the fire center is extinguished. The model also allows multiple fire centers, each of which can send out firelets, and it allows spotting if wind speeds are sufficient. In the latter, fire centers arise downwind of the existing fire (Clarke *et al.*, 1994). The paper by Clarke *et al.* (1994) provides a more detailed description.

9.5.2 Model spatial domain

The domain for the example model is a mountainous area near Santa Barbara, California. The area is represented by a 473 row by 395 column raster with 30 m resolution (total area = 16 815 ha). Elevation in the model domain ranges from sea level to 1209 m, representing significant relief so that topographic effects are clearly visualized. Spatial layers are all rasters stored in .GIF format and projected into UTM coordinates.

9.6   Summary

Fire is an excellent example of a phenomenon that lends itself to our general model of ecological transport (Ch. 2). Transport is initiated by the interaction of acute incidents, like lightning strikes, with chronic conditions, like fuel accumulation. Fire spread and spatial heterogeneity are tightly connected and give rise to one another. "Diffusion" from point sources, aggregation of multiple sources, and complex geometric configurations arise as the fire grows, coalesces, and advances horizontally and vertically. Finally, direct and indirect effects of fire are realized at places temporally and spatially removed from the initiating events and themselves serve to initiate other vectors of transport.

Fluvial transport

Eventually, all things merge into one, and a river runs through it. The river was cut by the world's great flood and runs over rocks from the basement of time. On some of the rocks are timeless raindrops. Under the rocks are the words, and some of the words are theirs. I am haunted by waters.

Norman Maclean (1976)

10.1 Transport system description

Water, often rich in suspended or dissolved substances, moves in response to gravity into and through porous media, over impermeable or saturated surfaces, and down channels (Fig. 10.1). In lakes and oceans, water transports dissolved and suspended substances in response to wind, Coriolis force, and gravity. Although both terrestrial and aquatic transport by water are intertwined and ecologically important, in this chapter we will limit our discussion to instances of fluvial transport in terrestrial systems.

One way to organize the formidable diversity included in this topic is to use the general transport scheme described in Part I of this book. Transport by water begins with an initiating event or chronic condition, moves an entity from one place to another, and results in effects or consequences. Initiating events include rainstorms, snowmelt, temporary subsurface drainage, and point or non-point sources of contaminant input. Permanent spring seeps, streamflow down channels, and groundwater movement through aquifers are examples of chronic conditions contributing to fluvial transport. Transported entities, moving along with water itself include materials (i.e. sediment and/or dissolved or undissolved chemicals), energy (i.e. the erosive potential of water), and/or information (i.e. genetic material in waterborne seeds) entities. The water vector moves these entities by mechanisms including bulk flow (advection), dispersion, and turbulent transport. Effects range from erosion to deposition, nutrient enrichment to chemical contamination, and habitat destruction to enhancement of plant and animal communities, among many others.

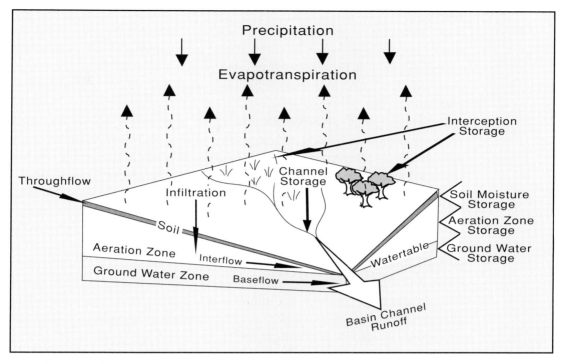

FIGURE 10.1
A schematic showing the most important components of the hydrologic cycle.
(Based on Freeze and Cherry; 1979; Fig. 1.1.)

10.2 Underlying principles for mode of transport

10.2.1 Principles of hydrology

We present here a few basic principles as examples of processes important for understanding transport by water, and we touch on some of the complexities that make modeling of this vector engaging. Discussions by Freeze and Cherry (1979), Ward and Elliot (1995), Thompson (1999), and many others offer more details.

Inputs, infiltration, and percolation
Ultimately, water inputs at the earth's surface arise from precipitation – either rainfall or snowfall – although the rerouting of water by humans for agriculture (irrigation) is significant in some places that would otherwise be arid. The variable frequency and magnitude of precipitation contributes to the heterogeneity of hydrologic inputs on land, which, in turn, affects transport processes, including rates of infiltration and runoff. Vegetation characteristics (e.g. canopy density), soil properties (e.g. porosity), and terrain features (e.g. slope, aspect)

also contribute to the variable rates by which precipitation infiltrates the soil, flows downhill, or re-evaporates (Schlesinger, 1997). Precipitation falling as snow can be redistributed by wind, stored until snowmelt occurs, or sublimated: all affecting the nature of hydrologic input (Fig. 10.2). Consideration of these inputs is a first step towards understanding and modeling fluvial transport across space.

Infiltration rates are determined by many land-surface properties. Vegetation intercepts raindrops, reducing their energy and their ability to erode soil on impact. Vegetation canopies also store precipitation, which either reaches the soil surface more slowly than it would otherwise or is re-evaporated (Schlesinger, 1997). Slower input rates allow more water to infiltrate the soil, resulting in less runoff. Vegetation also increases surface roughness, decreasing the speed with which runoff occurs (Abrahams et al., 1994). Plant roots, burrowing animals, and other soil organisms affect the infiltration rate by providing pathways into the soil for water flow (Beven and Germann, 1982). Soil properties, particularly porosity, also affect the rate of infiltration (Fig. 10.3). Other soil properties, like particle size (e.g. clay versus sand), affect infiltration rates through both their effect on porosity and through differences in their water adsorption potentials (Schlesinger, 1997).

Once water enters the soil, it is either stored, if the soil is below its field capacity (water holding capacity), or percolates downward in response to gravity. Field capacity is controlled by soil porosity (Fig 10.3), particle size, and composition. Clays, for example, have a higher ability than sands to hold water against the pull of gravity, both because pore spaces in clayey soils are narrower, contributing capillary resistance, and because of adsorption of water on the clay surfaces.

In saturated soils, the flow of water in response to gravity is described by Darcy's law (Eq. 10.1), originally formulated in 1856 by a French hydraulic engineer named Henry Darcy to describe water flow through sand-filled pipes. With modification, Darcy's law is also used for unsaturated flow through soils (Freeze and Cherry, 1979). It relates the flux of water (volume/time), or discharge (Q), to a driving force (the hydraulic gradient) and a conductivity (hydraulic conductivity) resulting from soil properties. One form of Darcy's law is:

$$Q = -K(\mathrm{d}h/\mathrm{d}l)A \tag{10.1}$$

where K is the hydraulic conductivity of a soil or other porous media, $\mathrm{d}h/\mathrm{d}l$ is the hydraulic gradient, and A is the cross-sectional area through which the flow is measured (Freeze and Cherry, 1979) (Fig. 10.4). Hydraulic conductivity is an empirically determined property of the porous medium (Schlesinger, 1997). Water that percolates through the soil contributes to base flow, which eventually reaches streams or becomes stored in groundwater.

FIGURE 10.2
Snowfall can be redistributed by wind to create zones of deposition where the wind
interacts with topography. Meltwater from these deposition zones is transported
across terrain by gravity, affecting vegetation patterns and the distribution of
waterborne entities.

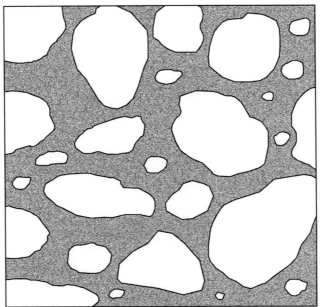

FIGURE 10.3
A schematic illustrating soil porosity. Water moves through the interstitial space between soil particles. Materials carried with the water undergo dispersion as they move around and among solid particles.

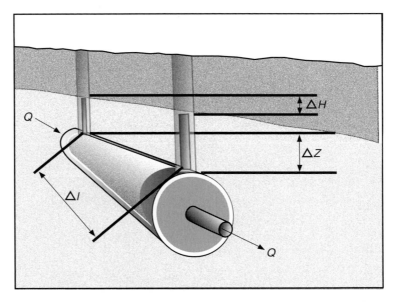

FIGURE 10.4
A schematic showing the components of Darcy's law ($Q = -K(\mathrm{d}h/\mathrm{d}l)A$). The discharge ($Q$) is determined by the change in hydraulic head (H) over some length (l) the hydraulic conductivity (K), and the diameter of the "pipe" through which the water moves. K is a property of the porous medium. (Based on Freeze and Cherry, 1979; Fig. 2.1.)

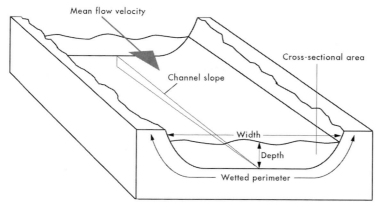

FIGURE 10.5
An idealized stream channel cross-section showing the components of the Chezy and Manning equations.

Runoff and overland flow

Once soil becomes saturated, excess water accumulates in surface depressions and, if inputs continue, begins to flow downslope as overland flow. The portion of rainfall or snowmelt flowing over the ground surface (overland flow) or through soils (interflow) to reach surface water is called runoff (Ward and Elliot, 1995). Runoff along the surface takes several forms, including unconfined sheet flow (interrill flow) and confined flow in channels (rilling, gullying, and piping) (Selby, 1993). We discuss overland flow in the context of erosion in Section 10.2.2 and general principles of channel flow in the following section.

Channel flow

As runoff volume increases or as slopes steepen, unconfined sheet wash coalesces into channels, which over time can become large (gullies) and connect with stream and river networks, which are themselves channels. Channels are complicated by heterogeneity in the environments that they traverse, but there are general principles describing channel flow that provide a starting point for modeling associated transport. In channels, the gravitational potential energy of water is converted to kinetic energy as the water moves down the channel or to heat energy (by friction) as it interacts with the boundaries of the channel. Two equations (Eqs. 10.2 and 10.3) are typically used to predict the velocity of water moving in a *uniform* channel (Hemond and Fechner-Levy, 2000). Both relate water velocity (V) to the slope and hydraulic radius (the ratio of cross-sectional area to "wetted perimeter") of the channel (Fig. 10.5). The first of these, the Chezy equation, is:

$$V = C(RS)^{1/2}, \tag{10.2}$$

where C is called the Chezy friction coefficient, R is the hydraulic radius, and S is the slope of the water surface.

The second, the Manning equation, is:

$$V = (1.49 R^{2/3} S^{1/2})/n, \tag{10.3}$$

where n is the Manning roughness coefficient and R and S are the same as in the Chezy equation. The C or n coefficients from the two equations are usually estimated empirically (Hemond and Fechner-Levy, 2000). For purposes of environmental transport, these equations represent two simple approaches for predicting movement of, for example, a chemical spill down a channel as a function of time. In natural systems, channels are rarely "uniform" and contaminant inputs (initiating events) have different duration. Chapra (1997) described temporal loading scenarios (i.e. impulse, step, linear, exponential, and sinusoidal), each of which affects transport differently.

While the Chezy and Manning equations offer starting points for predicting the downstream velocity of materials in a channel, dispersion and turbulent diffusion cause complex mixing of constituents entrained in the flow. Mixing complicates the picture by influencing both the time required for an entity to reach a location and the concentration of that entity in the water. Other complications arise through adsorption of contaminants onto materials in the bed or banks of the channel or onto suspended sediment (Axtmann and Luoma, 1991; Hemond and Fechner-Levy, 2000). Additionally, the Chezy and Manning equations in themselves offer no explicit means for dealing with spatial heterogeneity in stream channel characteristics. This can be partly resolved by distributing the C or n values down the channel and solving the equations iteratively.

Dispersion and turbulent transport

Dispersion is the movement of substances along concentration gradients resulting from the near-random motion of the fluid in which they are carried (Hemond and Fechner-Levy, 2000). It is analogous to pure molecular diffusion (Ch. 6) and is similarly described as a type of Fickian transport, but with a dispersion coefficient substituted for the diffusion coefficient described in Ch. 6. Freeze and Cherry (1979) defined mechanical (or hydraulic) dispersion as spreading owing to the motion of water during advection (bulk flow). Mechanical dispersion occurs in both groundwater and surface waters but the mechanisms, while analogous, usually operate at different scales in different media.

In groundwater, mechanical dispersion is best viewed as a microscopic process, occurring as water moves through pore spaces between solid particles (Fig. 10.6). According to Freeze and Cherry (1979), dispersion in groundwater is caused by three mechanisms. The first is the drag at the edges of pore channels created by the friction of water against solid walls. The second is differences in

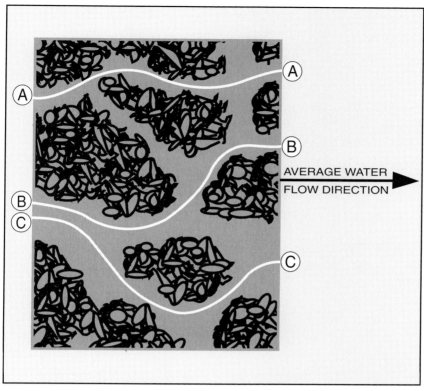

FIGURE 10.6
Mechanical dispersion occurs in porous media as material moves through pore spaces and around solid obstructions. Substances carried in the water become mixed as a result of this movement.

flow velocity caused by moving water encountering pore channels of variable width in porous media. The last mechanism is mixing caused by flow around obstructions. Tortuosity, an index of the path length a particle traverses as it moves around these solid particles, is one way to measure dispersion by this mechanism (see also Ch. 6).

In surface waters, both dispersion and turbulent diffusion cause mixing of suspended or dissolved entities. Because of friction, water closer to the bed or bank of a river generally moves more slowly than water near the center of the channel (Fig. 10.7). These velocity differences result in eddy formation and dispersion (Hemond and Fechner-Levy, 2000). Turbulent diffusion, though analogous, is mixing caused by the random motions of water interacting with channel irregularities and with the dispersive eddies. Both of these mechanisms result in random mixing and dilution of entities carried by the water. Contaminant plumes tend to elongate downstream as a result of velocity shear (dispersive eddies), and to widen downstream of their source as a result of dispersion.

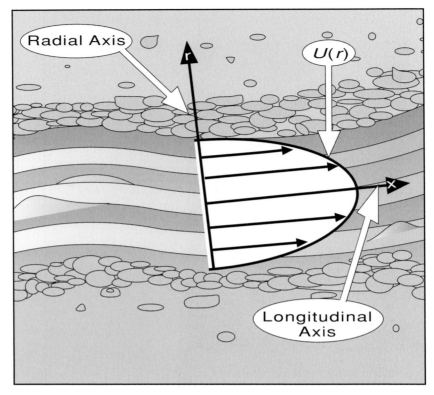

FIGURE 10.7
Friction with channel edges causes lower water velocities near these edges than in the center of the channel. These velocity differences cause eddies to form, which themselves result in dispersion of material carried in the stream. Turbulent diffusion also occurs as a result of the interaction of flow with irregularities in the channel.

10.2.2 Principles of erosion

Erosion by water is the detachment of materials at one location and their transport to another, where deposition occurs (Summerfield, 1991). One can think of erosion within the framework of an initiating event that detaches material, a transport mechanism that moves and eventually deposits it, and an effect caused by its removal or deposition. The characteristics of the land surface (or of underwater surfaces in the case of erosion in streams, lakes, and oceans) affects the rate and pattern of erosion. Erosion mechanisms operate at different scales and with different results. Locally and regionally, interrill and rill (channels small enough to be obliterated by plowing) erosion, gully erosion, and erosion by streams and rivers are examples (Fig. 10.8). At the continental scale, rates of erosion are complex functions of tectonic uplift, climate-driven

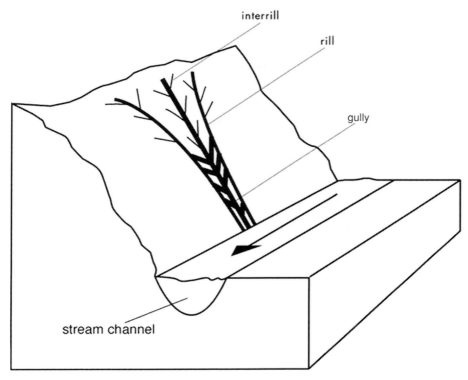

FIGURE 10.8
Gully, rill, and interrill erosion on a slope above a stream channel. Interrill erosion
is caused by shallow overland flow and raindrops. Interrills coalesce to form rills
(microchannels a few centimeters wide and deep); these themselves join to form
gullies under favorable slope conditions. Gullies are too large for obliteration by
tillage.

weathering and erosion (transport), and the accumulation of overlying weathered products or sediment (Stallard, 1992, 2000).

Interrill and rill erosion
Interrill erosion is erosion caused by the energy of raindrops impacting a surface
and by shallow overland flow (Selby, 1993; Ward and Elliot, 1995) (Fig. 10.8).
Raindrop splash can move significant amounts of soil over distances of a meter or
more but serves primarily to detach sediment from the soil, making it available
for other transport mechanisms, like shallow overland flow, that operate over
longer distances. According to Ward and Elliot (1995), who cited Watson and
Laflen (1986) and Liebenow *et al.* (1990), interrill erosion is a function of soil
properties, rainfall intensity, slope, and, sometimes, runoff rate. Other factors,
like the presence of plant cover or crop residues, can reduce erosion by absorbing
some of the energy of raindrops before they impact the soil.

When water flowing over a surface becomes concentrated in rills, it can cause rill erosion by detaching and carrying soil particles downslope (Ward and Elliot, 1995; Fig. 10.8). Rill erosion is a function of rainfall intensity and soil infiltration, and it usually dominates on longer and/or steeper slopes than does interrill erosion. The "Water Erosion Prediction Project" (WEPP) (Lane and Nearing, 1989; see discussion of modeling techniques below) included rill and interrill erosion processes in computer models. Moore and Burch (1986a) have provided a treatment of the physical processes of sheet and rill erosion, and other researchers (e.g. Abrahams *et al.*, 1996) have discussed rill erosion in non-agricultural settings.

Gully erosion

Gullies are erosional channels too large to be obliterated by tillage (Fig. 10.8; see also Fig. 10.14, below). According to Ward and Elliot (1995), gullies account for less sediment movement than do rill and overland flow but are a greater problem for farmers because they present obvious difficulties for plowing. Gully erosion rates depend on broader-scale factors than those causing rill erosion. Drainage area, soil characteristics, channel slope, and the orientation of the gullies all contribute. De Ploey (1990) presented a mathematical model describing conditions leading to precursors of gullies. Gulley creation begins when a critical shear stress is surpassed. At this shear stress threshold, erosion becomes non-selective, and soil particle movement occurs regardless of grain size.

Channel erosion

Stream channel erosion is caused by water flowing down banks (above the water level of the stream), by scouring and undercutting of the stream banks below the water surface, and by removal of material from the bed of the stream itself (Ward and Elliot, 1995). Land-use practices adjacent to the stream bank, particularly the removal of vegetation by grazing or by plowing too close to the stream, can influence the rate of bank erosion (Fig. 10.9). Mechanical alterations of stream geometry, like channel straightening, can also increase bank erosion above and scour erosion below the water level (Jones, 1997).

10.2.3 Principles of groundwater flow

Groundwater is usually defined as subsurface water found beneath the water table in soils and in fully saturated geologic formations (Freeze and Cherry, 1979; Fig. 10.10). To understand the flow of groundwater and its transport of dissolved constituents requires a broad understanding of chemistry, physics, geology, hydrology, and engineering. Its three-dimensional flow is often treated as a field phenomenon, analogous to electrical or magnetic

FIGURE 10.9
Stream-bank erosion exacerbated by heavy grazing along the Laramie River in
southern Wyoming.

fields. While Darcy's law (Eq. 10.1) is a starting point for understanding and
modeling groundwater flow, the complexity of mathematically characterizing three-dimensional heterogeneous systems quickly complicates the simple
relationship between driving force and resistance (Anderson, 1995). Because
of this complexity and because groundwater flow is such a multidisciplinary
phenomenon, we refrain here from presenting a process-based discussion and
instead refer interested readers to the classic groundwater text by Freeze and
Cherry (1979). Other excellent discussions of groundwater transport and modeling are found in Anderson and Woessner (1992), El-Kadi (1995), Spitz and
Moreno (1996), and Appelo and Postma (1996). The importance of the groundwater resource to humans, and the potential for its depletion or contamination,
continues to stimulate research and the growth of an already rich literature.

10.3   Modeling techniques

Fluvial models have several features: they (i) simulate transport initiated
by a variety of acute or chronic initiating events; (ii) describe movement of

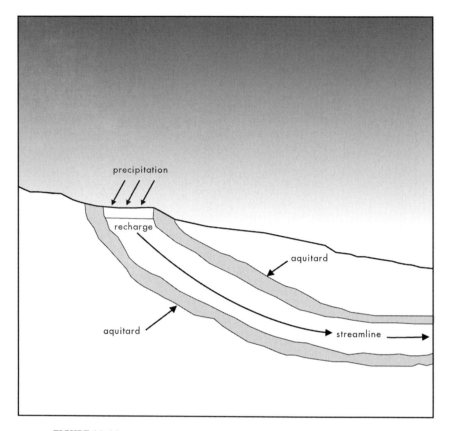

FIGURE 10.10
An aquifer contained between two relatively impermeable "aquitards." Recharge
from precipitation occurs where the porous formation forming the aquifer
intersects the surface.

specific entities by water; (iii) concern themselves with the mechanisms by
which the water vector carries entities; and (iv) predict the type and magnitude of the cause or effect resulting from the transport of the entities across
time and space. In this section, we discuss a few examples of fluvial models
selected from a much larger collection. Modeling of transport by water is a
well-developed field, and a comprehensive survey is impossible here.

One of the important issues in fluvial modeling is the distinction between
"lumped" and "distributed" models. Lumped models aggregate input parameters or output quantities over multiple model elements (e.g. raster cells), while
distributed models assign independent inputs and outputs to each cell. Development of fully distributed catchment or groundwater models is an active
research area (Anderson, 1995; Abbot and Refsgaard, 1996) but many models
described as distributed are in reality at least partially lumped (Beven, 1996).
Historically, lumping was necessary because of computer limitations. More

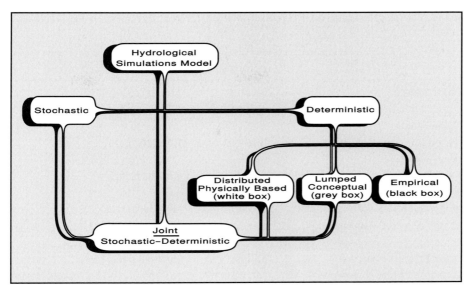

FIGURE 10.11
A simple classification of hydrologic models. (Modified from Abbott and Refsgaard, 1996.)

recently, as computational power has increased exponentially, problems with parameterization of complex models have taken precedence (Anderson, 1995). Assigning relevant properties to fine-scale grid elements across watersheds is difficult or impossible. Consequently, modelers in the real world are often forced to lump spatially distributed input parameters. One of the challenges for fluvial modelers is in developing methods for measuring spatially distributed properties of catchments at fine scales and incorporating them into the models (Anderson, 1995).

10.3.1 Types of fluvial model

Because hydrologic transport is ubiquitous in nature and ecologically and anthropogenically crucial, there is an extensive heritage of hydrologic models. Refsgaard (1996) has offered a useful classification of hydrologic models, which we summarize here (Fig. 10.11). A description of model types by Thompson (1999) is also useful and offers additional insights into classifying hydrologic models. Table 10.1 lists hydrologic transport models and acronyms used in the text.

Thompson (1999) divided hydrologic models at the broadest level into scale (analogue) models and mathematical (usually computer) models. Scale models are actual physical replicas of hydrologic systems, such as layers of sand in an aquarium into which dye can be injected to observe dispersion patterns.

TABLE 10.1. *Hydrologic transport models discussed in this chapter with their acronyms (sorted alphabetically by acronym)*[a]

Model name	Acronym	References
Agricultural Non Point Source	AGNPS	Young *et al.* (1989)
Areal Nonpoint Source Watershed Environmental Response Simulation Model	ANSWERS	Beasley and Huggins (1982)
Better Assessment Science Integrating Point and Nonpoint Sources	BASINS	EPA (2001)
Chemicals, Runoff and Erosion from Agricultural Management Systems	CREAMS	Knisel (1980)
Distributed Hydrology–Soil–Vegetation Model	DHSVM	Wigmosta *et al.* (1994)
Generation and analysis of model simulation Scenarios	GenScn	Kittle *et al.* (1998)
Hydrologic Simulation Program-Fortran	HSPF	Bicknell *et al.* (1993)
Modular three-dimensional finite-difference ground-water flow model	MODFLOW	McDonald and Harbaugh (1988)
Parameter estimating version of the modular model	MODFLOWP	Hill (1990)
Modified Universal Soil Loss Equation	MUSLE	Williams (1975)
The Precipitation Runoff Modeling System	PRMS	Leavesley *et al.* (1983)
The Enhanced Stream Water Quality Model	QUAL2E	EPA (1995)
Revised Universal Soil Loss Equation	RUSLE	Renard *et al.* (1991)
Systeme Hydrologique European	SHE	Abbot *et al.* (1986)
Solute Transport with Advection, Resuspension and Settling	STARS	Jakeman *et al.* (1998)
Soil and Water Assessment Tool	SWAT	Arnold *et al.* (1993)
A hillslope hydrology simulator	TOPMODEL	Wigmosta *et al.* (1994)
Universal Soil Loss Equation	USLE	Wischmeier and Smith (1978)
The Water Erosion and Prediction Project	WEPP	Lane and Nearing (1989)

[a] References are either primary or contain useful descriptions of the associated model.

Although important, we will not discuss these models further because they do not lend themselves to spatially explicit modeling. Mathematical models form the core of present-day hydrologic modeling and were divided by Refsgaard (1996) into deterministic and stochastic models.

Deterministic models produce the same output when presented repeatedly with identical input. They were divided by Refsgaard (1996) into three main groups (Fig. 10.11). The first empirical models (often called "black box" models) were built from analyses of the relationship between input and output time series. An example, the unit hydrograph model, converts rainfall input into streamflow output without consideration of the processes linking the two (Thompson, 1999). Statistical techniques have been used to create mathematically more advanced hydrologic models than simple empirical models, but they still use a "black box" approach. The second group of deterministic models includes "conceptual models," which are, according to Refsgaard, a kind of book-keeping system that continuously accounts for moisture content and storage. Spatially, conceptual models are almost always lumped, with parameters and variables averaged over catchments or parts of catchments. The third type of deterministic model described by Refsgaard are physically based models. These models usually require partial differential equations to represent a continuum, and they consider hydrologic *processes* at least in part, although "significant deficiencies still persist at the level of the process descriptions" (Refsgaard, 1996). Additionally, physically based models are often at least partly spatially distributed.

Stochastic models are derived from time-series analyses of historical records and can generate hypothetical sequences of events possessing the same statistical properties as the time series used to develop the model (Refsgaard, 1996). These models frequently use Monte Carlo techniques to generate statistical distributions, and they usually do not explicitly consider process. Stochastic models are another type of "black box" model (Fig. 10.11). Combining elements of deterministic and stochastic models may be the best way to represent our current understanding of hydrologic process (deterministic) and to recognize our inability to provide accurately spatially distributed parameters and input variables (Refsgaard, 1996). Typically, a deterministic core operates within a stochastic frame in models combining the two approaches.

Model dimensionality is another character that can help to distinguish among models at all levels of Refsgaard's scheme (Thompson, 1999). One-dimensional models are sufficient for some stream runoff simulations (e.g. PRMS, QUAL2E), while for non-point source models two dimensions may be required (e.g. ANSWERS). Prediction of groundwater transport, especially in thick or steeply sloping aquifers, requires three dimensions (Spitz and Moreno, 1996).

Finally, the type of event simulated by the model is an important consideration. Event models, like rainfall-runoff models, simulate a stream hydrograph after a short-term rainfall event. Continuous models, by comparison, simulate the hydrologic system as a whole and update system components at regular time steps (Thompson, 1999). Continuous models can operate over longer time scales (i.e. months, years) than can event-based models.

Because modeling of transport by water is such a diverse and well-developed field, we must limit our discussion here to examples from each of several areas of research with strong ecological implications. **Hydrologic modeling** concerns itself with the prediction of water quantity, where water is both the transport vector and the transported entity, and with the hydrologic cycle in general.

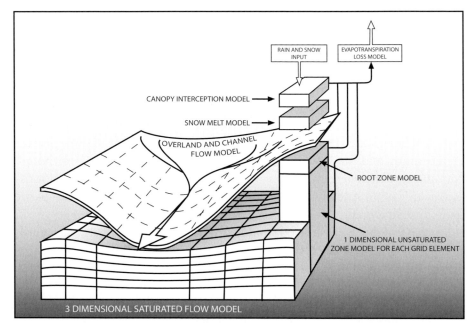

FIGURE 10.12
Schematic showing the SHE model components. (Modified from Refsgaard, 1996.)

Erosion modeling is concerned primarily with the transport and deposition of soil, at least at ecologically meaningful time scales. **Water-quality modeling** simulates transport and changes in concentration of dissolved and suspended nutrients and contaminants. **Groundwater modeling** treats the movement of water below the water table in soils or in saturated aquifers. Of course, there are linkages between all of these and any model may consider contributions from several categories.

10.3.2 Hydrologic modeling

All of the models discussed in this chapter (Table 10.1) could be broadly classified as hydrologic models; however, for our purposes, we define hydrologic models in a narrower sense as models predicting the quantity or timing of water flow or general elements of the hydrologic cycle. In this sense, hydrologic models correspond to the "hydrologic submodels" imbedded in modeling systems like PRMS (Leavesley *et al.*, 1983) or SHE (Abbot *et al.*, 1986) (Table 10.1). PRMS simulates general basin hydrology and SHE simulates the hydrologic cycle (Fig. 10.12). While these broad hydrologic systems are clearly linked to specific types of transport (i.e. erosion, sediment transport, or solute transport in groundwater), we treat those topics separately.

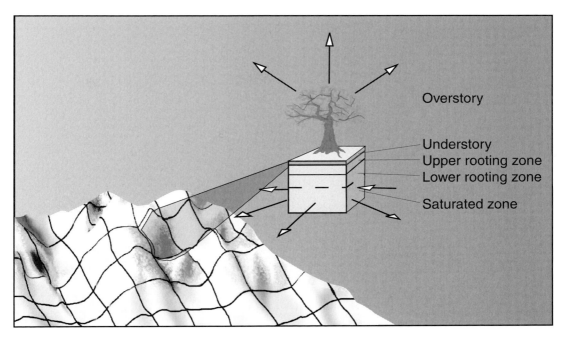

FIGURE 10.13
Schematic showing the structure of the DHSVM model. Energy balance equations
are solved for each grid cell. Grid cells exchange subsurface flows with neighboring
cells. (Based on Wigmosta *et al.*, 1994.)

Hydrologic responses to climatic events were traditionally predicted using
empirical tables, charts, and graphs relating the intensity and duration of inputs
(e.g. precipitation) to the timing and magnitude of outputs (e.g. stream discharge) (Ward and Elliot, 1995). While useful, these "black box" models do
not explicitly consider process and have limited application in spatially heterogeneous domains.

More recently, modelers have adopted process-oriented, spatially distributed
approaches to hydrologic modeling. PRMS, for example, is a quasi-process-
based model to allow exploration of the effects of precipitation, snowmelt intensities, and land-use practices on water balance, flow regimes, flood peaks and
volumes, soil–water relationships, sediment yields, and groundwater recharge
in basins (Leavesley *et al.*, 1983). The model uses a combination of physical
processes and empirical relationships in a spatially distributed setting.

Another example of a distributed, process-based hydrologic model is the
DHSVM (Wigmosta *et al.*, 1994), which calculates runoff from catchments
and considers plant canopy interception, transpiration, evaporation, and snow
accumulation and melt (Fig. 10.13). In the model, topography, represented by a
digital elevation model (DEM), exerts controls on incoming solar radiation,
air temperature, precipitation, and downslope water movement. Vegetation is

FIGURE 10.14
Gullies are channels that may grow from year to year as a result of headcutting or
lateral erosion. This gully is eroding a colluvial fan in Death Valley, California.

represented by a two-layer canopy model that controls evapotranspiration and
a two-layer root-zone model. Snow accumulation and melt are determined
by an energy-balance model. A "quasi three-dimensional" saturated subsurface flow model handles below-ground water movement and links grid cells
hydrologically. DHSVM operates on daily or subdaily time steps. Importantly,
". . . downslope redistribution of soil moisture via the saturated subsurface
transport scheme is explicit: that is, accounting is done on a pixel-by-pixel
basis, as contrasted with the statistical representation used, for instance, by
TOPMODEL" (Wigmosta *et al.*, 1994).

10.3.3 Erosion modeling

Background and history
Erosion of soil by wind or water is of enormous economic importance in agricultural lands and has stimulated a long history of models to predict erosion
rates (Williams *et al.*, 1996b) (Fig. 10.14). The complexity of the erosion process,

however, was daunting to early investigators. Culling, in a 1960 paper, lamented that "it will prove to be too ambitious an undertaking at present to construct a theory from fundamental physical principles, whether at the hydrodynamic or hydraulics level." Hergarten and Neugebauer (1996) sounded even less optimistic and stated: "Trying to integrate all these processes into a physical model seems to be hopeless." Despite the discouraging preludes, these and other researchers forged ahead enthusiastically into the realm of erosion modeling and have generated an enormous body of literature and a variety of empirical and quasi-process-based models. Foster (1991) provided a concise review of historical erosion modeling. More comprehensive treatments are found in Penning de Vries *et al.* (1998).

At the most general level, erosion modeling can be divided into those models that take an empirical approach, relating measurements of properties in the field to erosion rates, and those that rely on mathematical descriptions of physical processes. Because of their great practical utility, empirical models have been used extensively in agricultural management and are still used today. Process models, however, have the potential to offer more robust depictions of spatially heterogeneous environments in real-world catchments.

Empirical erosion models

Empirical equations relating erosion rate to measurable characteristics of land surfaces formed the core of erosion modeling for over 60 years. A. W. Zingg published an equation in 1940 predicting rill and sheet erosion as a function of slope length and steepness. Zingg's equation was eventually improved by adding expressions accounting for climate, cropping, and land-management practices (Smith and Whitt, 1957). These advances set the stage for development of USLE (Wischmeier and Smith, 1978), arguably the most popular empirical model for estimating soil erosion by water. USLE is still used to identify potentially erodable land and to suggest erosion management strategies (Foster, 1991). It is imbedded whole or in part within many modern computer erosion models, for example CREAMS (Knisel, 1980), SWAT (Arnold *et al.*, 1993), AGNPS (Young *et al.*, 1989), and ANSWERS (Beasley and Huggins, 1982). USLE has also been modified to improve its performance, for example in MUSLE (Williams, 1975) and RUSLE (Renard *et al.*, 1991) (Table 10.1).

USLE relates soil loss to empirical factors and takes the form:

$$A = RKLSCP, \tag{10.4}$$

where R is a rainfall and runoff erosivity index, K is a soil erodability factor, L is a slope length factor, S is a slope steepness factor, C is a cropping management factor, and P is a conservation practice factor. In the equation, the index R represents the driving force for erosion, and the other factors represent resistances.

All of the factors in the USLE equation are empirical and must be estimated in the field or collected from existing databases (Wilson, 1996).

Process-based erosion models

The empirical erosion model USLE has a number of shortfalls, including lack of ability to predict sediment deposition, the assumption of uniform slopes, and treatment of runoff as uniform over a catchment (Wilson, 1996). In situations where slope steepness is variable (Nearing, 1997), or where heterogeneity is present in other relevant factors, the model falters. Examples in agricultural settings include ridge and contour tillage and strip cropping (Foster, 1991). Natural systems are almost always spatially variable. By the late 1960s, Meyer and Wischmeier (1969) had begun to apply physical principles rather than empirical factors in order to improve the ability to deal with heterogeneity.

Laflen *et al.* (1991) described the history of WEPP (Lane and Nearing, 1989) (Table 10.1), initiated by the US Department of Agriculture in 1985 with the goal of creating a flexible, process-based modeling system. This effort, spanning 10 years, resulted in a set of models that were better able to handle heterogeneity at the subcatchment level. WEPP considers three basic processes critical to erosion: detachment, transport, and deposition. By dividing catchments into subcatchments or grid cells, some spatial variability could be captured, although much of the modeling in WEPP was still "lumped."

According to Foster (1991), concern for non-point-source pollution motivated development of other process-based models including ANSWERS and CREAMS (Knisel, 1980; Beasley and Huggins, 1982) (Table 10.1). Some of these models (e.g. CREAMS) used a combination of empirical relationships and process-based equations, but all recognized some degree of spatial heterogeneity within watersheds.

The ANSWERS model (Beasley and Huggins, 1982) is an example of a spatially distributed model incorporating elements of empirical and process-based approaches. ANSWERS simulates surface runoff and erosion in predominantly agricultural watersheds (Wilson, 1996) by dividing them into raster cells and routing flow from cell to cell to the watershed outlet. The model considers detachment of soil particles by raindrops, as well as detachment and transport by overland flow.

ANSWERS combines elements of empirical modeling from USLE with erosion processes. For each grid cell, the model considers topography (elevation, slope, aspect), soils (porosity, moisture content, field capacity, infiltration capacity, and the USLE *K* factor), land cover (percentage cover, interception, the USLE *CP* factor, surface roughness, and retention), channel characteristics (width, roughness), and rainfall (De Roo *et al.*, 1989; Wilson, 1996). Rain falling on each cell results in interception, infiltration, surface storage, surface flow, subsurface flow and sediment detachment, transport, and deposition. The output from each cell is routed to adjacent downstream cells.

The use of geographic information systems in erosion modeling
GIS has been linked to many erosion prediction models, including USLE and ANSWERS, especially as a tool for handling spatially distributed input data. GIS is useful for managing complex inputs from heterogeneous catchments and for visualizing model results. Model "front ends," like GenScn (Kittle *et al.*, 1998; Table 10.1) provide graphical user interfaces to automate the process of organizing input and scenario data for fluvial models in general.

10.3.4 Water quality modeling

Background and history
Water quality modeling concerns itself with the input, transport, and effects of materials dissolved or entrained in moving water, and with their concentration through time and space. Often these materials are associated with sediment, suggesting a link between erosion modeling, which concerns sediment entrainment, and water quality modeling. The movement of pollutants and contaminants are of critical concern to humans and ecosystems. Chapra (1997) provided a comprehensive overview of surface water quality modeling. He described four major periods of modeling evolution, spanning the last 70 years. During the first phase, from 1925 to 1960, computer processing was either unavailable or in its infancy, and models were restricted to linear kinetics, simple geometries, and steady-state receiving waters. Modeling emphasis was on dissolved oxygen (e.g. Streeter and Phelps, 1925) and bacteria, and models were one dimensional.

The second period of water quality model development corresponded to the advent of digital computing. While dissolved oxygen was still the focus of research, computers allowed researchers to consider more complex geometries, kinetics, and time-varying simulations. Models were extended from one to two dimensions (i.e. "wide" estuaries and bays) and focus was shifted from small domains to entire drainage basins. Modeling emphasis remained on point sources of contaminants.

During the 1970s, societal awareness and a booming economy in the USA led to a shift of focus from point sources to general environmental concerns, including non-point sources of pollution and eutrophication (Chapra, 1997). Representations of biological processes became more mechanistic, and complex nutrient/food chain models were developed (e.g. Chen, 1970; Canale *et al.*, 1976).

The most recent phase of model development, from 1977 to the present, has focused on toxic substances, acid rain, and recognition of the role of solid matter in transport (e.g. Chapra and Reckhow, 1983; O'Connor, 1988). The association of toxicants with suspended particles, the view of the food chain as a concentrator of toxicants, and advances in sediment–water interactions and hydrodynamics all played a role. Further increases in computer speed and availability during this period allowed advances in graphical user interfaces, decision-support systems, and the extension of models from two to three dimensions.

The list of water quality models that have been developed is long, and its literature is extensive. In the following sections, we cite examples of models that address important aspects of water quality modeling, or that represent unique approaches to simulation. Interested readers should refer to Chapra (1997) and many others for more comprehensive treatments.

Sediment transport modeling

The interaction of pollutants with solid matter suspended in water has motivated an interest in sediment transport modeling for over 20 years (Anderson, 1954, 1981; Pickup *et al.*, 1983; Axtmann and Luoma, 1991), although from ecologic and geomorphologic perspectives, the effects of sediment movement are more far-reaching than contaminant fate alone (Fig. 10.15). Sediment transport models are usually designed to simulate transport resulting from one or more initiating events by several possible transport mechanisms. Primary consequences, among others, include deposition of material, transport of adsorbed contaminants, changes in stream energy, and channel evolution. Examples of secondary consequences range from changes in riparian plant communities (Naimen and Bilby, 1998) to alteration of fish spawning patterns (Everest *et al.*, 1987; Benda *et al.*, 1992).

Sediment transport is initiated by events that either introduce sediment to moving water (e.g. erosion, biotic activity, atmospheric deposition) or resuspend sediments that have previously settled (e.g. turbulence, mechanical disturbance). Erosion by ice, wind, water, and chemical weathering results in the input of inorganic allochthonous material: material of mineral origin that is usually denser than organic sediments. Organic material includes dead cells, bacteria, and aquatic organisms. Sediments are found either suspended in the water or residing temporarily at or near the bottom of water bodies. Suspended sediments are always in transport. Bottom sediments can be resuspended and transported. Many of the erosion models discussed above (e.g. ANSWERS, CREAMS; Table 10.1) provide estimates of sediment inputs to streams from land.

Mechanisms for the transport of sediment in water include advective (bulk) flow, turbulent transport, and Fickian mixing (dispersion) (Fig. 10.16). As sediments move, they settle through gravity (Hemond and Fechner-Levy, 2000). A mathematical relationship known as Stokes' law (Eq. 8.1; p. 138), discussed in the context of wind transport, relates settling rates to particle properties

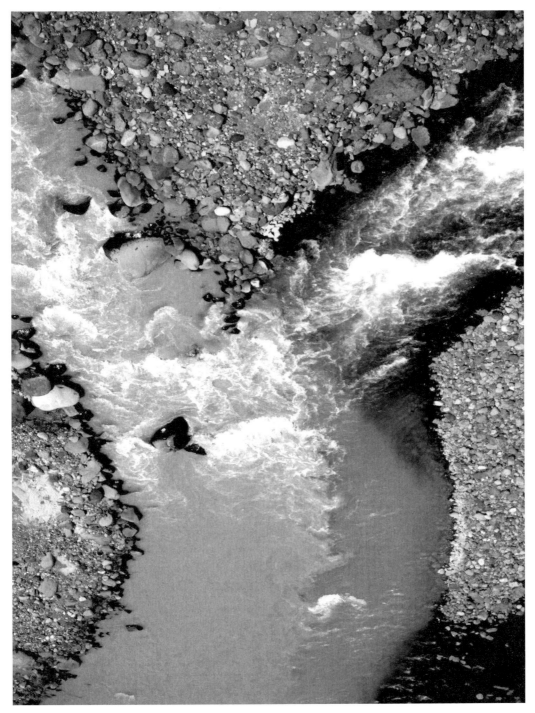

FIGURE 10.15
The confluence of the Rio Sucio (Dirty River) in Costa Rica with a smaller
clear-flowing creek. Rivers transport huge amounts of sediment with profound
effects on downstream systems.

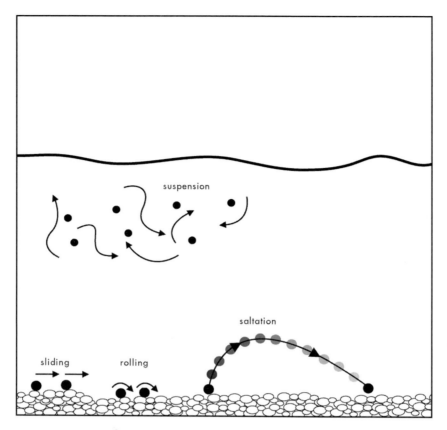

FIGURE 10.16
Various mechanisms that result in sediment transport. Particles in suspension are small and light enough to remain suspended by the turbulent eddies in the flowing water. Heavier particles can saltate – moving along the bottom in short "jumps" as they are broken free of the streambed by turbulence but settle again due to gravity as they move downstream. Still heavier particles move along the bed by rolling or sliding.

(particle shape, diameter, density) and properties of the water (density, dynamic viscosity). Chapra (1997) suggested that most sediment transport models use empirically derived settling rates because in natural systems the Stokes' law assumptions do not hold. Since flow is turbulent, many organisms are buoyant, and clumping of sediment is common. Sediment movement is also quite variable in space and time (Jones, 1997). For these reasons, most sediment transport models are based on empirical relationships rather than on physical process.

Water quality models often include a sediment transport component. De Vries (1993) presented mathematical formulae describing several types of sediment transport in rivers but suggested that the assumption necessary for these equations to be valid (steady uniform flow) is usually not met. Jakeman *et al.*

(1998) described a model called STARS (Table 10.1) that helps to identify sources and sinks of sediment within stream reaches. STARS accounts for mass conservation, advection, settling, and resuspension mechanisms. The model is parameterized empirically from measurements of sediment concentration. QUAL2E (EPA, 1995) uses advective and dispersive mass transport to simulate transport of dissolved constituents and sediment in one-dimensional stream reaches. The model uses a mass-balance approach and is solved for steady-state conditions. Other water quality models also simulate sediment loads. More details on sediment transport for interested readers are found in Bagnold (1988), Yang (1995), and Chadwick and Morfett (1998).

Point-source models

Point sources of pollutants are "discrete, localized and often readily measurable discharges of chemicals" (Hemond and Fechner-Levy, 2000; Fig. 10.17). Examples include pipes discharging industrial or domestic waste or storm water, leaking underground storage tanks, and localized chemical spills on or near a surface water body. Viessman et al. (1989) suggested that point-source water quality models are usually restricted to confined water bodies, like channels or groundwater, and generally deal with continuous inputs of contaminants. Point-source models are frequently based on the Streeter–Phelps equations (Streeter and Phelps, 1925), originally developed for simulating dissolved oxygen concentrations in streams receiving sewage (Chapra, 1997). An example of a model that simulates the fate of both point-source and non-point-source pollution is BASINS (Table 10.1), developed by the US Environmental Protection Agency (EPA) in 1996. BASINS has been linked to ArcView (ESRI, Redlands, California) to enable handling of spatial data input and output. Point-source contamination in BASINS is simulated by a one-dimensional routine called Toxiroute, which considers stream reach data and first-order pollutant decay.

Non-point-source models

Non-point pollution sources include agricultural runoff carrying fertilizer and/or pesticides, urban runoff contaminated with surface pollutants, or pollution from numerous point sources considered together (Hemond and Fechner-Levy, 2000). Non-point-source pollution is more difficult to measure than that from point sources, because it is, by definition, spatially distributed. Viessman et al. (1989) described many non-point-source models as "loading models," which actually trace the transport of pollutants from their origin until they reach a water body, where further transport is passed to water quality models (Thompson, 1999).

Quenzer (1998) classified non-point-source loading models as constant concentration, spreadsheet, statistical, rating curve or regression, and buildup/

FIGURE 10.17
Point source of contaminant entering a stream in Laramie, Wyoming.

washoff models, depending on the mechanism used in the model. Most of these methods are empirical, with the exception of the buildup/washoff models, which can employ explicit representations of process. Examples of non-point-source models include HSPF (Bicknell *et al.*, 1993), CREAMS (Knisel, 1980), BASINS (EPA, 2001), and ANSWERS (Beasley and Huggins, 1982) (Table 10.1).

10.3.5 Groundwater modeling

Groundwater flow and transport is fundamentally three dimensional, a circumstance that adds modeling complexity to an already complex phenomenon (Fig. 10.10). While horizontal groundwater movement is typically orders of magnitude more rapid than vertical movement (Day, 2000), both are important in the subsurface environment. Anderson (1995) has surveyed the history of groundwater modeling in the twentieth century, and we follow her discussion here. She credited Darcy's work on water flowing through sand, and his formulation of Darcy's law (Eq. 10.1, p. 184), published in 1856, as the foundation of modern groundwater modeling. Analogue models based on electrical fields and capitalizing on the notion of groundwater as a field phenomenon also provided early insights into the nature of groundwater. Theis (1935) noticed similarities between heat and groundwater flux and developed models of transient groundwater flow to wells based on the heat flow literature (Anderson, 1995). His models were mathematical but, like analogue models, drew on the analogies between the phenomena.

With the increasing availability of computers in the 1960s, numerical models became standard, and development of computer models of groundwater flow and transport is a robust field of research today (Anderson and Woessner, 1992; Anderson, 1995; El-Kadi, 1995). Bachmat and Bear (1964) used a Fickian analogue (Ch. 6) to model dispersive groundwater transport, substituting a dispersion coefficient for the diffusion coefficient. Measurement of the dispersion coefficient, however, especially in heterogeneous media, is difficult. Theis (1967), by anticipating the modeling complexities imposed by three-dimensional heterogeneity in aquifers, and Freeze (1975), by emphasizing uncertainty in groundwater flow, stimulated research into stochastic and geostatistical methods, which led to the 1977 Penrose Conference. This conference laid the groundwork for more recent groundwater transport modeling (Anderson, 1995).

The three-dimensional heterogeneity of aquifers and the difficulty of measuring parameters used in groundwater transport models are fundamental difficulties that modelers must overcome. Methods to accomplish this range from trial and error "tuning" of parameters until model output matches measured groundwater properties, to parameter estimation models that take a more systematic approach (Anderson, 1995). An example of the latter is MODFLOWP (Hill, 1990), developed to estimate parameters for the MODFLOW model (McDonald and Harbaugh, 1988), which itself has become an industry standard for groundwater modeling (Anderson, 1995). Another approach is the use of stochastic methods, including use of "effective parameters," geostatistics, Monte Carlo techniques, and conditional simulation (Anderson, 1995). All of these methods use random hydraulic conductivity fields built around statistical distributions representing the modeled aquifer. In short, explicit representation of the complex, three-dimensional, geologic heterogeneity in aquifers is extremely difficult and continues to challenge groundwater modelers today. Readers are referred to El-Kadi (1995) and Anderson and Woessner (1992) for detailed discussions of these issues.

10.3.6 Scale and resolution

The spatial and temporal resolution of model inputs is a crucial issue for model development and sensitivity and deserves special mention. Temporal resolution primarily affects the handling of water inputs (precipitation events, snowmelt, etc.). Spatial resolution is concerned with depictions of terrain and surface properties (Fig. 10.18). While many models use spatially averaged or

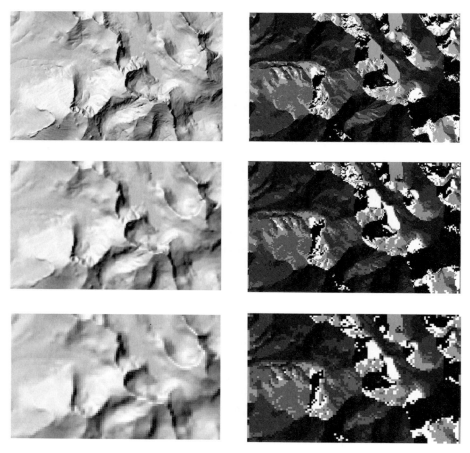

FIGURE 10.18
As the spatial resolution of digital elevation models (left column) decreases (top to
bottom), calculated flow routes (right column) change. The resolution of elevation
data used as input to hydrologic models can have profound effects on model output.

"lumped" inputs to sidestep heterogeneity and computational limitations, spatially distributed models are becoming more common. The effect of spatial resolution on hydrologic model design and performance has become a research area
in its own right, especially with regard to terrain analysis (e.g. Moore and Burch,
1986b; Moore et al., 1991; Costa-Cabral and Burges, 1994). Moore et al. (1991)
provided an overview of DEMs and problems encountered when resolving terrain for hydrologic modeling. In particular, they discussed the calculation of
drainage networks from DEMs and problems caused by artificial "flat spots" in
elevation data. Costa-Cabral and Burges (1994) examined the effect of terrain
data on computation of areas contributing to flow at particular points. They
noted that flow is sensitive to the data structure (e.g. rectangular grids versus
points) used to describe the surface.

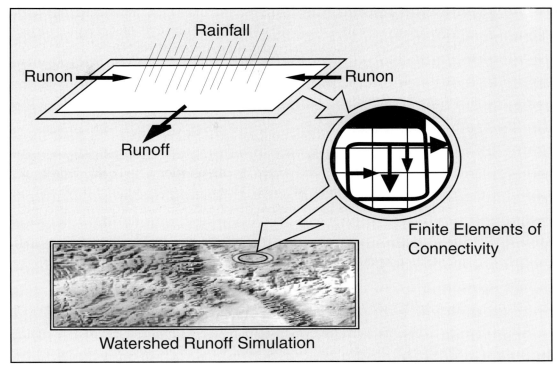

Rainfall

Runon

Runon

Runoff

Finite Elements of
Connectivity

Watershed Runoff Simulation

FIGURE 10.19
Schematic showing the basic structure of Arc.water.fea. (Modified from Vieux *et al.*,
2001; unpublished manual.)

10.4 Introduction to the example model

The example model for this chapter is an ArcView extension called
Arc.water.fea, originally developed in the C programming language for the
US Army Corps of Engineers and ported to a PC/NT environment and ArcView
by Vieux (2001) at the University of Oklahoma. The model supplied on the CD
is a demonstration version. A description of the model is available in Vieux
and Gaur (1994) and Vieux (2001). The full version of the model can be used to
delineate watersheds, extract data, simulate river-basin hydrology, and display
model outputs as hydrographs and maps of water depth. The demonstration
version of the model supplied here is described by Vieux *et al.* (2001, unpublished manual).

10.4.1 Principles of the model

Arc.water.fea is a finite element model based on a raster depiction of
watersheds (Fig. 10.19). The model domain is described by a DEM from which

the model delineates basins, streams and drainages, accumulates flow, and derives slopes (Vieux *et al.*, 2001; unpublished manual). Model inputs include these DEM-derived parameters, spatially distributed infiltration rates based on soil data, and hydrologic roughness derived from land-cover data. Rainfall data can vary in space and time and come from radar data, rain-gauge data, or other sources. All inputs are presented to the model as raster data sets with identical pixel size and location.

In the model, overland flow from one cell to another follows a map of drainage direction derived from the DEM. Overland flow is calculated using the kinematic wave analogy, which is based on conservation of mass and momentum equations driven by slope steepness (Vieux *et al.*, 2001; unpublished manual). The mathematical methods are described in the demonstration manual (Vieux *et al.*, 2001) and by Vieux and Gaur (1994). The modeling techniques work best where channel and land-surface slopes are steep, and they may not be appropriate for flow through relatively flat terrain.

10.4.2 Explanation of the spatial domain

The Arc.water.fea demonstration model supplied on the CD is parameterized for the 2400 km^2 Illinois River Basin in northeastern Oklahoma and western Arkansas. DEM data for the basin are provided at a spatial resolution of 960 m. The demonstration simulates flow from the basin during a storm event that occurred on 8–11 June, 1995.

10.5 Summary

The distribution of water is critical in terrestrial ecosystems, both because of the importance of water itself for biota and because of its ability to transport dissolved and suspended materials. Geomorphologically, water plays a central role in shaping the surface of the land over time. Because water is pervasive on our planet, and because it is important for so many physical and biological processes, it is difficult to describe its transport in just a few pages. In this chapter, we have touched on some of the simple relationships between the flow of water on and below the land surface, and the physical factors that drive and resist these flows. We have also introduced a few of the types of fluvial transport emphasized by modelers, and the broad strategies they use to simulate these complex hydrologic systems. Even now, with a considerable heritage of modeling to draw from, consideration of spatial and temporal heterogeneity in hydrologic models is an active area of research, but one that is likely to yield important insights into ecological systems.

11

Animal movement

A journey of a thousand miles must begin with a single step.

Lao-tzu (604–531 BC)

11.1 Transport system description

In its larval stage, *Danaus plexippus* (the North American monarch butterfly) is interested primarily in eating. The desire to mate comes later, with its famous transition from larva to butterfly. Humans are fascinated with butterflies, perhaps because their life cycle is a metaphor for the emergence of beauty from apparent mediocrity or maybe because of the analogy to the human condition, which hovers butterfly-like between a desire to eat and an urge to mate. Grace (1997, p. 8) described it as follows. "Like Dr. Jekyll and Mr. Hyde, the monarch butterfly and caterpillar show two contrasting sides of the same life. One is an elegant and gregarious beauty with a penchant for courtship and travel, the other, a stay-at-home glutton with voracious habits and a solitary disposition." However, the monarch is also an example of animal movement across spatial and temporal scales and a starting point for thinking about this mode of transport.

A monarch caterpillar, born from an egg, moves within a small territory, eating exclusively from milkweed plants, storing energy, growing, molting, and avoiding predation, in part through its diet of toxic milkweed steroids. At first its tiny size (~2 mm) limits the caterpillar to browsing on leaf hairs in the immediate vicinity of its egg, but it soon grows enough to consume leaf blade material from its "home" leaf and other leaves on the home plant (Urquhart, 1987). Detection of potential predators (like birds) can cause caterpillars to curl into balls and drop onto the ground, resuming random movement in search of a new host plant only when the danger has passed (Grace, 1997).

After about three gustatory weeks, the caterpillar molts a final time, producing a chrysalis in which it metamorphoses into a monarch butterfly. The butterfly dries its wings and flies (at speeds up to 50 km/h), feeding on flower nectar, water, and fruit juice; it expands its spatial domain further and builds strength

through the summer months. Finally, falling autumn temperatures stimulate southward migration to overwintering grounds in California or southern
Mexico. Grace (1997, p. 47) described this:

> The monarchs sweep south in flurries like advancing weather systems,
> cyclones of butterflies heading for the tropics to merge and build into a
> hurricane of hundreds of millions of beating wings over Mexico. Taking
> advantage of rising warm air currents, they may circle to altitudes as
> high as 1200 metres (4000 feet) before gliding down to catch the next
> wave up. They are returning to their ancient homeland, for monarchs
> are really tropical butterflies and have expanded their range.

In the spring, the monarchs "remigrate" north to their breeding grounds, where
the cycle begins again. From the slow, local movement of the newborn caterpillar
traversing the plane of a leaf surface to the long distance three-dimensional
flight of the migrating butterfly, *D. plexippus* is a model for animal transport of
matter, energy, and information across scales and through time.

Most animals possess some means of locomotion: they move about in space
across the surface of the land, through the air, or in water. Importantly, as
animals move they serve as vectors, transporting, sometimes inadvertently,
energy, matter, and information. Their movement, initiated at one place in
space and time, causes an effect at another. The vector of animal movement is
unique because it is not a purely physical process as are other transport vectors.
Animal movement is a *behavior* and animals, either consciously or unconsciously,
must *decide* to move. As a consequence, a dependable deterministic prediction
of animal movement in space is fundamentally impossible. This is reflected in
a variety of stochastic and probabilistic animal movement models.

11.2   Underlying principles for mode of transport

The study of animal movement in ecology has been approached from
the perspectives of foraging behavior (Stephens and Krebs, 1986), population dynamics (MacArthur and Wilson, 1967; Dunning *et al.*, 1995), dispersal
(Lidicker and Caldwell, 1982; Wiens, 2001c), and home range utilization (Burt,
1943; Worton, 1987) to name a few. Ims (1995) challenged that:

> Unfortunately, neither theory nor empirical knowledge on animal
> movement patterns are currently sufficiently developed to serve as a
> mechanistic basis for landscape ecology. Elements of relevant theory on
> the subject are scattered over several approaches, each favored by
> different scientists with very different backgrounds; from ethology
> (optimal search theory), through behavioral ecology (optimal foraging)
> and population ecology (habitat selection and dispersal theory) to
> mathematics and computer modeling (e.g. diffusion models).

Ims (1995) advocated combining elements of the different theories in order to predict more effectively animal movement in spatially heterogeneous domains and across scales.

The consideration of heterogeneity is arguably more important for animal movement than for many of the other transport vectors. Animals often move *because of* heterogeneity in their environments: they search actively for patches of suitable habitat or relative richness of food supply among less optimal terrain. Wiens (2001c) expanded on this:

> Dispersal is usually considered in terms of movement from one location and the consequences of arriving at other locations. Movement between these locations is often viewed as a linear process, on which the characteristics of the intervening area have little effect. The area between the beginning and ending of dispersal, however, is not a featureless matrix of unsuitable habitat, but is instead a richly textured mosaic of patches of different shapes, sizes, arrangements, and qualities: a landscape.

Animal transport is initiated by physiological cues, like the perception of hunger; by external stimuli, like the attack of a predator or a change in the weather; or by inadvertent occurrences, like the attachment of a seed to the wool hiking sock of a passing mammal. Transported entities include matter (e.g. nutrients, soil, food items), energy (e.g. metabolic heat, latent heat, kinetic energy), and information (e.g. genetic material, visual or olfactory signals). Entities are carried by diverse modes of locomotion – walking, running, slithering, burrowing, jumping, flying, swimming, oozing – but other processes, like entity attachment or infection, are also important. Finally, effects of animal transport range from physical changes in the environment (e.g. soil compaction) to structural changes in ecosystems (e.g. redistribution of populations) to subtle social interactions among and within species (e.g. territorial disputes).

All aspects of animal movement are important for understanding transport, but we can only touch on a few of them here. We introduce a few approaches used to study animal movement and direct interested readers to more detailed treatments. Turchin (1998) has a comprehensive survey of the analysis of animal movement in the framework of population redistribution. Dunning *et al.* (1995) reviewed spatially explicit population models in a special issue of *Ecological Applications* devoted to that topic. Baker (1978) and Lidicker and Caldwell (1982) discussed aspects of dispersal and migration. Foraging, with an emphasis on optimal foraging theory, has been reviewed by Stephens and Krebs (1986). Okubo and Levin (2001) discussed animal movement in the context of diffusion theory.

11.2.1 Initiating events: stimulus and motivation

The initiation of transport by animals is perhaps best discussed in the context of motivation, implying both a stimulus to action and a decision to respond to that stimulus. Motivating stimuli can be internal or external to the animal and are born from physiological necessity, genetic predisposition, or, arguably, whimsy (i.e. animal "play"). These stimuli can be short-duration events, like the explosion of a lioness from the cover of grass into a herd of zebra, or chronic conditions, like changes in forage quality resulting from extended drought in the Sahel. Lidicker and Caldwell (1982) emphasized that, while environmental conditions are clearly important, individual characteristics of the organism must also be considered, including age, sex, physiological state, and genetic composition. We discuss some important modes of animal movement in this section in the context of motivation, recognizing that motivation initiates movement.

Taxis (response to stimulus)

The term *taxis* was defined by Campbell and Reece (2001) as "movement oriented towards or away from some stimulus." Although this definition describes most animal movement, taxis is usually restricted to a fine-scale response to light, heat, gravity, or chemical concentration. Turchin (1998), in a discussion of modeling taxis, even included "preytaxis," which is the movement of animals towards their prey. These movements are called "positive" when they are oriented toward the stimulus and "negative" otherwise (Fig. 11.1). Examples include the flight of moths towards light (positive phototaxis), the movement of fish away from toxic substances (negative chemotaxis), and the movement of a snake into a sunny patch to absorb the heat of the sun (positive thermotaxis). In these cases, environmental heterogeneity is an essential motivator as animals proceed toward or away from areas where a particular condition is found.

Foraging

It is the fortunate animal that can stay in one place and await the arrival of food. While some creatures set traps for their prey (e.g. spiders, ant lions) or wait for prey to wander into their area of influence (e.g. sea cucumbers, mollusks), most must continuously search to find enough food to survive. Foraging for food is one of the primary forces motivating animals to move and, as a result, influencing their territory size (and hence their transport domain). For example, Greenwood and Swingland (1983) described work by Gill and Wolf (1975) relating the territory size for the golden-winged sunbird (*Nectorinia reichenowi*) to flower density. For this bird, territory size was not fixed but depended upon the presence of enough *Leonotis nepetifolia* flowers to provide the nectar on which it fed.

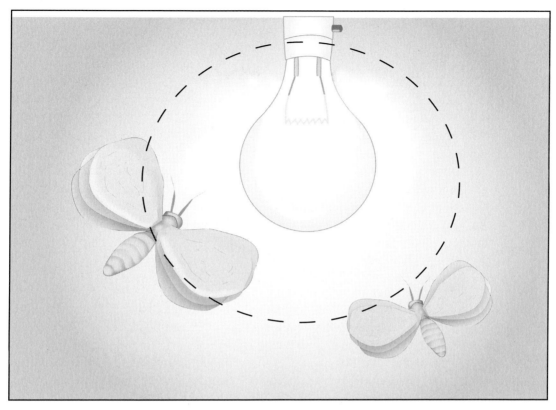

FIGURE 11.1
Taxis is the movement of an organism towards or away from some stimulus. This
graphic illustrates positive phototaxis; the movement of a moth towards a light.

Foraging strategies are the subject of much research (Pyke, 1983; Stephens
and Krebs, 1986) and considerable controversy (Pierce and Ollason, 1987). One
paradigm, called optimal foraging, is the idea that animals forage to maximize their fitness – the success with which they reproduce (Fig. 11.2). Optimal
foraging theory (MacArthur and Pianka, 1966; Stephens and Krebs, 1986)
explains foraging choices as the optimization of the rate of energy gain enjoyed
by a foraging animal, where energy gain is used as a surrogate for the hard-to-
measure ideal of future reproductive success (Pyke, 1983; Cuthill and Houston,
1997). Foraging animals must decide when to move or cease moving, where
to move, and how to move (e.g. whether to walk or run). Optimal foraging
theory says that these decisions are made to maximize energy uptake rates
(Pyke, 1983).

Complications arise when other variables, like predators and social interactions, are considered. The monarch caterpillar maximizes leaf intake on
a single plant *until* the arrival of a potential predator stimulates it to drop

FIGURE 11.2
Optimal foraging behavior balances the energy required to find new, potentially rich
sources of food by moving between patches and energy savings realized by foraging
locally within a patch for food that may be of lower quality.

onto the ground in defense and eventually search for another milkweed plant.
Additionally, foraging decisions may be influenced by scale. In heterogeneous
landscapes, foragers may make decisions differently about within-patch foraging than about between-patch foraging, although optimization of energy
uptake remains the implicit goal (Pyke, 1983; Ims, 1995).

   A problem with optimal foraging theory is that the behavior of foraging by
animals in many cases can be explained without it (Pyke, 1983). Pyke suggested that, although qualitative evidence is common, precise quantitative
support for optimal foraging is lacking. This, however, may be more of an
experimental problem than a shortcoming of the theory, and optimal foraging has been used successfully to explain animal behavior in many cases
(Kamil *et al.*, 1987). Stephens and Krebs (1986) mentioned alternative theories,
including random foraging models, satisficing, and descriptive approaches.
Satisficing includes satisfaction of minimum requirements and involves

decision-making when information or time is limited. Ward and Blaustein (1992) suggested that satisficing may be particularly useful for explaining some foraging behavior.

Regardless of how one approaches foraging, it is clear that energy uptake by feeding is one of the primary factors motivating animals and initiating movement and transport by them.

Dispersal

Another context for looking at animal movement is the study of dispersal, associated with a variety of motivations. The most parsimonious definition of animal dispersal requires only the departure of an individual from its "living place" (Lidicker and Caldwell, 1982). A dispersing animal makes the decision to leave an area that it has previously occupied, and then it travels some distance away. Lidicker and Caldwell (1982) acknowledged that animal movements span a continuous range of scales, and they noted that dispersal includes "accidental or idiosyncratic dispersal on one extreme, through various kinds of adaptive dispersal, to round-trip types of migration indulged by entire populations on the other extreme." The monarch anecdote illustrates this temporal and spatial dependency. One could argue that dispersal occurs when the young caterpillar journeys from the "home" leaf on which it was born to another leaf on the same milkweed plant. Alternatively, the plant might be considered "home" and dispersal occurs when the caterpillar drops onto the ground in response to a bird and moves to a new plant. Or the entire milkweed patch and its surroundings could be the home territory and departure for the Mexican highlands in autumn constitutes dispersal. The point is that explicit consideration of spatial and temporal scale is critical for understanding dispersive movements.

What are the motivations that cause animals to disperse? Lidicker and Caldwell (1982) noted that, while we can observe dispersal events, it is often difficult to infer the proximal motivating factors. They suggested that animals must balance (whether consciously or unconsciously) the advantages and disadvantages of departing from their home area (Fig. 11.3). Wiens (2001c) echoed this, noting that dispersal "determines the probability that an individual currently 'here' will later be 'there', and as a consequence be exposed to different opportunities and risks." Potential advantages can be environmental, such as escape from unfavorable conditions or reduced exposure to predation, competition, and disease; or they can be genetic, such as increased mate selection and avoidance of inbreeding. Disadvantages accrue through the uncertainties of achieving these benefits, the energy cost of travel, increased exposure to predation during travel, and genetic consequences such as founder effects. At some level, whether conscious or intuitive, an animal weighs these advantages and disadvantages when it "decides" to disperse.

FIGURE 11.3
Fish migrating upstream to spawn risk being eaten along the way, a risk that they would not suffer if they did not migrate.

Dispersal is frequently studied in the context of population dynamics. Turchin (1998) provided a comprehensive discussion of animal movement and movement modeling in this context, and Dunning *et al.* (1995) reviewed spatially explicit population models. We refer interested readers to these discussions rather than compressing such a rich literature into a few paragraphs. Detailed discussions of other aspects of dispersal are provided by Wolfenbarger (1946) (small organisms), Lidicker and Caldwell (1982) (general theory), Horn (1983) (general theory), McCleave *et al.* (1984) (fish dispersal), Kot *et al.* (1996) (invading organisms), and Okubo and Levin (2001) (diffusion theory).

Migration
The southward flight in autumn of monarch butterflies from the USA and Canada to their Mexican wintering grounds, followed by their return in the spring, is a spectacular example of insect migration. The sheer number of butterflies involved, representing a large portion of the monarch population, attracts human interest because of the clustering of so many colorful creatures. Other migratory events, like the return of the American robin in the spring, or the movement of wildebeest in the Serengeti (Fig. 11.4) in response to the dry season, evoke the passing of time and the rhythms of nature.

The term migration is fraught with definitional problems, in part because it depends on scale, which is relative and continuous, and in part because researchers do not agree on motivation. Rogers (1983) contended that a dichotomy between behavioral and ecological definitions is a false division

FIGURE 11.4
Wildebeest (*Connochaetes taurinus*) following seasonal rains migrate from Tanzania into Masai Mara National Park, Kenya.

between the related questions "why?" and "how?" In the context of insect migration, Johnson (1969) suggested that migration involves an adaptive *change* in breeding habitat. However, many examples of animal behavior that are commonly called migration (e.g. salmon spawning) include the return of an animal to the *same* breeding grounds. Baker (1978; p. 5) suggested that "for most people, certain conspicuous components of these movement patterns, such as the return to the original habitat, geographical directionality, periodicity, and the large distances involved, qualify these movements as migrations." Lidicker and Caldwell (1982) argued that migration "is dispersal plus an implied returning to the original area by an individual or its immediate descendants." We recognize that *migration* is necessarily a "fuzzy" concept because precise scalar limits cannot be placed on concepts like "home range" or "large distances."

Perhaps the most overriding basis for migration is evolutionary. For some species, selective pressures reward individuals that exhibit migratory behavior. An understanding of selective pressures is necessary for understanding the adaptive value of migration (Dingle, 1980). However, triggering mechanisms, including specific changes in weather conditions, photoperiod, or food availability, as well as social cues, like the movement of other animals, act in concert with this genetic imperative (Lidicker and Caldwell, 1982). Timing is also important, especially when future conditions (e.g. winter storms) have the potential to trap animals and prevent escape, or when simultaneous arrival at the migratory destination is important.

11.2.2 Transported entities

Matter, energy, and information are carried by moving animals either intrinsically (e.g. the mass of the animal, the heat produced by metabolism) or extrinsically (e.g. inadvertently attached pollen, seeds, or parasites). The diversity of animals and their habitats is enormous, as are the possibilities for transport. We discuss some important transported entities, while remembering that the categories are not mutually exclusive. Transport of pollen, for example, is movement of matter, energy, and information.

Matter

The most obvious transport accomplished by animals is the movement of the animals themselves. The bodies of monarch butterflies *en route* to Mexico are composed of materials that provide food for predators (usually birds) and detritus when they die. As metabolizing organisms, animals deposit part of the food consumed at one place as waste at another (Meyer and Schultz, 1985). Feces and urine are important sources of nutrients in some ecosystems, particularly where grazers forage in dense herds (McNaughton *et al.*, 1988, 1997; Frank *et al.*, 1994; Hobbs, 1996). Bird guano deposited on small oceanic islands is rich in nutrients, including phosphorus and nitrogen (Lindeboom, 1984; Sanchez-Pinero and Polis, 2000). In the tropics, where nutrients are tightly cycled, animal excrement is consumed with remarkable speed by creatures such as dung beetles. Antlers, shed annually by many ungulate species (e.g. elk, caribou, deer), are rich in phosphate and calcium and are scavenged by other animals.

Animals, seeds, and spores are also transported on moving animals. A study of the transport of animals and microbes as "passengers" on other animals (Jones *et al.*, 1998) illustrates how complex ecological relationships can influence the spread of disease. In eastern US forests, oak trees (*Quercus* spp.) produce large crops of acorns every two to five years and very few acorns in other years. Mast years (years with large acorn crops) attract both white-footed mice (*Peromyscus leucopus*) and white-tailed deer (*Odocoileus virginianus*) to feed on the acorns. Both of these species are primary hosts of black-legged ticks (*Ixodes scapularis*), which are vectors of the microorganism (*Borrelia burgdorferi*) that causes Lyme disease in humans. The mice are responsible for transmitting the Lyme disease bacteria to tick larvae, which, after reaching reproductive age, attach themselves to deer (Jones *et al.*, 1998). More infected ticks on wide-ranging deer lead to wider

FIGURE 11.5
A termite mound in Kenya. Termites control the temperature and humidity within
their mounds.

transport of Lyme disease and higher risk to humans who may encounter the
ticks. The interaction of the spatial and temporal heterogeneity of mast years
with the transport of bacteria and ticks by mice and deer weaves a net of initiating
events (mast years), animal transport, and spatially distributed effects (Lyme
disease spread).

Energy

Animals convert food materials into biomass and energy; each is transported as
they move. Latent and sensible heat are dispersed into the animal's immediate
environment, occasionally with significant ecological consequence. Termites,
for example, control the temperature and humidity within intricately designed
termite mounds (Korb and Linsenmair, 1999) (Fig. 11.5). A more dramatic example is the interaction of foraging giant hornets (*Vespa mandarinia japonica*) and
honeybees (*Apis cerana japonica*) (Ono *et al.*, 1995). Hornets, rallying around a
pheromone signal produced by a successful forager, attack the honeybee nest
en masse, but the honeybees, also alerted by the pheromone, defend vigorously.

Hornets captured in the fray are quickly surrounded by a dense ball of hundreds of honeybees, raising the temperature inside the ball to levels that are lethal for the hornets but not harmful to the bees (Ono *et al.*, 1995).

Of more obvious importance ecologically is the conversion of chemical energy into activities and behaviors. Animals walk, run, trample, burrow, fly, swim, eat, build nests, and reproduce, all as a result of stored energy carried with them as they engage in these activities. Energy derived from food consumed at one place enables animal activity at another place; this is the heart of the ecology of animal transport. Beavers use the chemical energy contained in the plant material they eat to move logs for damming streams. Ultimately this results in changes to the physical landscape with a suite of secondary effects. Activities like this are implicit in many of the contexts that we touch on in this chapter.

Animals also transport energy as the chemical constituents of their bodies, and as metabolic waste. Predators feed on prey because they are able to convert the food to energy, which is used for other activities (including finding more prey). Meyer and Schultz (1985) described the significant transport of nutrients and energy to coral colonies in marine environments by French and white grunts (*Haemulon flavolineatum* and *H. plumieri*). The grunts feed in seagrass beds at night and rest over the coral during the day, depositing feces, which increases the productivity of the coral. Meyer and Shultz (1985) suggested that the energy and nutrient inputs may also stimulate benthic communities beneath the coral.

Information

As animals move they transport information in their own genetic material, in "hitchhiking" propagules of other organisms, and via signals that may be broadcast either purposefully or inadvertantly. A deer moving through a forest, for example, carries its own sperm or eggs and may carry plant propogules attached to its fur. Additionally, the deer makes noise as it walks, signalling potential predators of a possible meal. All of these are forms of information. Other examples include pheromone trails, scents used by predators to find prey, territorial marking, and ritualistics movements that signal locations of food (e.g. the honey bee "waggle dance").

SEEDS AND SPORES

Genetic information in seeds and spores is carried from place to place as animals move, either by attachment to fur or feathers or in the animal's gut. Many plants and animals have evolved mutually beneficial relationships that increase the reproductive success of the plants while providing food or shelter to the animals

(Estrada and Fleming, 1986). Perhaps the most common example is the production of fleshy fruits to attract animals that act as seed dispersers. Animals gain nutrients from the fruits while carrying the indigestible seeds away from the parent plant. While there is a risk of seed damage by mastication or digestion, some seeds *require* disturbance of the seed coat (scarification) to germinate. Strawberry and raspberry seeds are examples; after ingestion by birds, the seed coats are weakened enough by digestion to improve germination, but not so much as to damage the plant embryo. Scarified seeds are then deposited with a small amount of fertilizer (feces), further increasing their odds of success.

Because flight is a potential mechanism for the long-distance transport of seeds, birds and bats have been studied more than other animals as seed transport vectors (Estrada and Fleming, 1986; Willson, 1993; Hickey, 1997; Hickey *et al.*, 1999). Birds can potentially carry seeds for long distances (many kilometers), although more typically the transport is local. Seeds usually have short residence times in bird digestive tracts, either because they are regurgitated soon after ingestion or because fruit pulp is passed quickly (van der Kloet, 1988; Hickey, 1997). Hickey (1997) suggested that frugivorous bat dispersal of seeds is usually local (hundreds of meters) because bats do not normally travel far from the parent plant. Charles-Dominique (1986) found that bats were important for secondary succession in tropical forests of French Guyana because they transport seeds of pioneer species (e.g. *Cecropia* spp.) into forest gaps. Even monarch butterflies have been implicated in plant dispersal. Van Zwaluwenberg (1942) noted the simultaneous establishment of monarch colonies and their food plants on Canton Island in the Pacific. Presumably, the monarchs were responsible for the establishment of the plants (Carlquist, 1981), although the mechanism for this was unclear. Clark *et al.* (1998) suggested that long-distance seed dispersal by any vector is rare but possible and describes the shape of the probability–distance curve as leptokurtic, with long-distance dispersal events forming the wide tails of the distribution.

Other vertebrates, and especially omnivorous mammals, can also act as vectors for seed dispersal (Willson, 1993; Hickey, 1997; Hickey *et al.*, 1999). Many animals eat fruits and pass viable seeds some distance from the parent plant. Hickey *et al.* (1999) argued that facultatively frugivorous carnivores are particularly likely to transport seeds over long distances, because their teeth are less liable to damage seeds than those of herbivores, their digestive tracts are relatively short, and because they wander widely in search of prey. They studied seed dispersal by American martens (*Martes americana*) on Chichagof Island in Alaska and found that blueberry (*Vaccinium* spp.) seeds isolated from marten scat had higher germination rates than seeds left within intact fruit, suggesting that martens can be effective seed dispersal vectors. Primates are also important seed dispersers (Estrada and Fleming, 1986). Estrada and Coates-Estrada

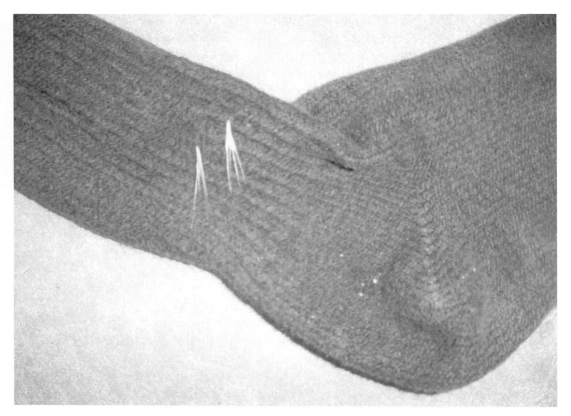

FIGURE 11.6
A cheatgrass (*Bromus tectorum*) seed attached to a sock. Many seeds are adapted to attach to passing animals to ensure dispersal.

(1986), for example, found that howling monkeys (*Aluoatta palliata*) were vectors for seed dispersal for 10 families of plants in humid tropical forests of Mexico. Janson *et al.* (1986) suggested that brown capuchin monkeys (*Cebus apella*) influence fruit characteristics because of selective pressure resulting from their dispersal of seeds.

Seeds can also disperse by becoming attached to animals either by chance or design. Researchers have successfully germinated seeds carried in mud on the feet of birds and in bird feathers (Carlquist, 1981). Many plants have evolved structures facilitating attachment of seeds to passing animals. Longspine sandbur (*Cenchrus longispinus*) and common cocklebur (*Xanthium strumarium*) produce burs that attach to animal skin and fur, and to human clothing (Whitson *et al.*, 1996). In the western USA, downy brome, or cheatgrass (*Bromus tectorum*), has spread into many native grasslands, in part because its needle-like seeds become easily attached to fur and clothing (Ogle, 2000) (Fig. 11.6).

Humans are especially effective seed dispersal agents, partly because we travel locally and globally and partly because we consciously carry seed or plant material from one place to another, for either cultivation or decoration (Vitousek *et al.*, 1997). Like other animals, we also transport seed accidentally. Roads and trails are recognized as source areas for the dispersal of exotic species by humans (Ogle, 2000). Williamson (1996) and many others review the biology of invasive species.

POLLEN

Once a monarch butterfly emerges from its cocoon, dries its wings, and takes flight, it becomes a vector for the dispersal of pollen. The monarch flies from one milkweed plant to the next, feeding on nectar that it draws from conspicuous flowers. Small, pollen-filled sacs, called pollinia, stick to the monarch's legs as it gathers the nectar and are deposited when the butterfly moves to a new flower, allowing cross-fertilization among the milkweed (Urquhart, 1987).

Transport of pollen by animals, particularly birds, bats, and insects, is essential for the reproduction of many plants and has been studied by humans for many years (e.g. Darwin, 1889) (Fig. 11.7). E. O. Wilson introduced a book on pollinators by Buchman and Nabhan (1996) with the following.

> There is a welded chain of causal events that leads directly to our species: if plants, including many food and forage crops, as well as natural floras, must have insects to exist, then human beings must have insects to exist. And not just one or two kinds of insects, such as the friendly and lovable honeybees, but lots of insect species, vast numbers of them. The reason is that millions of years of coevolution have finely tuned the relations between particular plants and their special pollinators. The shapes and colors of the flowers, their scent, their location on the stalks, the season, and daily schedule of their pollen and nectar offerings, as well as other qualities that we admire but seldom understand, are adjusted precisely to attract particular species of insects; and those specialists in turn, whether beetles, butterflies, bees, or some other group, are genetically adapted to respond to certain kinds of flowers. In lesser numbers the same is true of the interactions between plants and species of birds, bats, and other vertebrates dependent on diets of pollen and nectar.

This passage speaks to the fundamental importance of pollen transport by animals for the survival of life, and to the relationships that have developed to ensure that pollination occurs. Animal transport of the information encoded in pollen is a key process in most ecosystems.

FIGURE 11.7
A rufous hummingbird (*Selasphorus rufus*) feeding on fireweed (*Epilobium angustifolium*) nectar in the Medicine Bow Mountains of Wyoming. Hummingbirds inadvertently serve as pollinators when they transport pollen grains from one flower to another.

11.2.3 Effects of animal transport

Animal transport affects nearly every aspect of ecosystem structure and function – including abiotic and biotic components across scales from microscopic to global and momentary to evolutionary. Immigration, emigration, dispersal, predation, and the spread of disease are examples of changes in population structure resulting from animal transport. Herbivory, nutrient transport, pollination, seed dispersal, and evolutionary pressure are critical for structuring plant communities. Soil compaction, nutrient cycling, and greenhouse gas emission (e.g. methane) alter the abiotic environment. All of these, at least partly, result from transport of various entities by animals and all interact with one another in complex ways.

Modeling is one way to work towards understanding some of the processes and interactions that we have mentioned above. In the next section we highlight some of the techniques that have been used to build models of animal transport.

11.3 Modeling techniques

The rich history of modeling animal movement and the distribution of populations in space and time reflects the fundamental ecological importance of animal transport (Okubo and Levin, 2001). Turchin (1998) noted that "quantitative understanding of the consequences of movement for population dynamics is practically impossible without constructing and testing mathematical models." Turchin emphasized, however, that the utility of models would be severely compromised without a strong empirical basis for parameterization and model testing. Many animal movement models are based on field data (e.g. radio telemetry, mark–recapture) that identify animal locations at specific times. As we mentioned above, because animal movement includes stochastic and behavioral elements, accurate deterministic modeling is fundamentally impossible (Bovet and Benhamou, 1988).

11.3.1 Types of model

The range of applications of animal movement models complicates their classification. Many authors (e.g. Ims, 1995) have fitted models into schemes mixing model structure (e.g. random-walk models) and model context (e.g. foraging theory). Others (e.g. Turchin 1998) have described many model structures within the context of a single area of emphasis, like population dynamics. Turchin (1998) noted that "in addition to differing in their goals, movement models can be classified in many other ways: simple versus detailed, stochastic versus deterministic; and by their mathematical structure (whether key variables such as the population density, space, and time are discrete or continuous . . .)." He stressed the distinction between Lagrangian models, which are centered on the moving individuals, and Eulerian models, which are centered on the points in the space through which animals move. These distinctions really describe the frame of reference from which models are conceived, and the approaches are fundamentally interchangeable. Because we are interested in a broad view of animal transport modeling, we organize our discussion around model structure and give examples where possible from a variety of subdisciplines. We emphasize the approaches used to model movement and transport rather than the details of specific animal behaviors. While there are many more approaches and examples than we could discuss here, we touch on some that are widely used.

Scale models

Simplified mechanical representations of phenomena (scale models) have been described for several of the transport vectors discussed elsewhere in this book (e.g. wind, water). Scale models are arguably less intellectually satisfying than more theoretical approaches for viewing the behavior of natural systems, but they provide basic insights and, importantly, empirical data upon which more elegant solutions are based.

Animals have been visualized as "mechanisms" since the time of Descartes, as have attempts to simulate animal movement using mechanical devices. DeAngelis and Yeh (1984) described a mechanical animal (named *Machina speculatrix*) built by Walter (1951) to explore animal searching movement. *M. speculatrix* explored randomly until presented with obstacles around which it could maneuver (searching in heterogeneous space?). It also responded to light stimuli by preferentially approaching sources of light (taxis) unless the light was "too intense," in which case it would move away. While mechanical models of animals probably have limited application for studying animal behavior in complex natural systems, DeAngelis and Yeh (1984) noted that many of these mechanisms behave in ways that are "uncannily realistic," and that, like more complex mathematical models, they are really just abstractions of biological systems.

Conceptual models

Conceptual models are mental constructs that allow us to consider important aspects of phenomena while ignoring less-relevant details (Selby, 1993). These

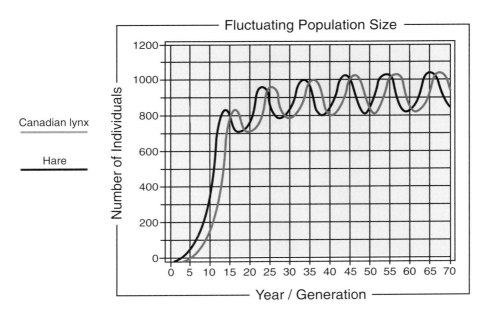

FIGURE 11.8
A conceptual model of the relationship between the populations of snowshoe hares and lynx in the Arctic. Lynx populations typically lag the population of snowshoe hares by a couple of years. The interaction between predator, prey, and food supply of both results in populations that vary sinusoidally over time.

constructs have been crucial for increasing our understanding of many aspects of animal activities (e.g. predator–prey relationships) (Fig. 11.8). Quantitative modeling of animal movement is often done within the framework of conceptual models that simplify general classes of phenomenon. Examples include optimal foraging theory (Stephens and Krebs, 1986), metapopulation dynamics (Hanski and Gilpin, 1991), predator–prey equations (Lotka, 1925; Volterra, 1926) and home-range concepts (Burt, 1943; Worton, 1987). While some of these, like the Lotka–Volterra predator–prey equations, *imply* movement in space (movement *must* occur for a predator to find and capture prey), they do not treat space explicitly. Others, like optimal foraging theory or home-range descriptions, although not always spatially explicit, come closer to a consideration of space. Optimal foraging is based on the notion that spatially distinct patches of food are scattered about the environmental domain in which an animal forages. Costs to the forager are incurred as it moves from patch to patch. Metapopulation dynamics depend upon the spatial arrangement of metapopulations, and transport between them. Many analytical models were developed to explore particular conceptual frameworks, just as experiments are designed to explore hypotheses.

Empirical models

Modeling the fundamentally non-deterministic phenomena of animal movement and transport is well suited to empirical approaches that quantify the relationship between observed animal behavior and metrics describing that behavior (e.g. density–distance relationships). In some cases (e.g. home-range descriptions), purely statistical models describe the distribution of animals in space. In others (e.g. random-walk models; see below), animal behavior is characterized by probability distributions that provide information for spatially explicit representation of simulated movement events. We cite two examples of empirical modeling approaches here, while recognizing that many others exist (see Turchin, 1998).

An animal's home range is defined intuitively as ". . . that area traversed by the individual in its normal activities of food gathering, mating, and caring for young" (Burt, 1943). In a review of home-range models for animal movement, Worton (1987) suggested that "the most useful methods for analyzing home range data sets involve describing an animal's range by a probability density function of location, called the utilization distribution." He gave a more rigorous definition of home range as "some fixed (usually 95%) confidence region obtained from the animal's utilization distribution function." Seaman and Powell (1996) described the utilization distribution as "a probabilistic model of home range describing the relative amount of time that an animal spends at any place."

Methods used to estimate the utilization distribution are reviewed by Worton (1987, 1989) and Seaman and Powell (1996). One popular non-parametric technique, called the kernel density estimator (Worton, 1989), estimates density in any number of dimensions (usually two for home-range studies) and is unaffected by grid size and placement (Seaman and Powell, 1996). Sequential locations of a single individual, single locations of different individuals, or measurements of properties other than location (e.g. plant biomass) are examples of observations that can be characterized using this tool. In the context of home-range studies, the density is an estimate of the amount of time an animal spends at a particular place, and consequently it is a tool for exploring animal preferences for specific locations. Seaman and Powell (1996) have given a detailed review of this method and alternative estimators.

A second example of statistical modeling tackles the problem of the spread of zebra mussels (*Dreissena polymorpha*) among freshwater lakes and rivers in Wisconsin, USA. Zebra mussels were introduced from Europe, probably via shipping channels during the last century; by 1990, they occurred in all of the Great Lakes and the Mississippi River, as well as in some of the many small freshwater lakes in the Great Lakes region (Buchan and Padilla, 1999). Zebra mussels are commonly carried to inland freshwater lakes on boats transported overland by recreational boaters. Buchan and Padilla (1999) reported that reaction–diffusion models (see discussion of diffusion modeling below) underestimated the spread rate of the mussels because they did not account for the relatively rare, long-distance dispersal events caused by boaters.

Surveys characterizing the pattern of recreational boat transport between lakes allowed Buchan and Padilla (1999) to estimate the annual spatial and temporal activities of recreational boaters over a large geographic area. They used this survey information to improve estimates of the zebra mussel spread rates. Because diffusion models are based on local, essentially random movements of dispersing entities, they fail to predict rare long-distance "jumps," like those caused by humans transporting mussels via roads. Statistical models like the one developed by Buchan and Padilla (1999) are one way to account for these rare events while maintaining a spatial context. These authors contended that such models are robust and can be applied to any invading species that experiences long-distance dispersal between "island" habitats.

Diffusion models

The analogy between the nearly random movement of animals and the diffusion of molecules or other particles (see Ch. 6) has led to a heritage of diffusive models of animal movement and population spread (Skellam, 1951; Okubo, 1980; Kareiva, 1982; Ims, 1995; Turchin, 1998; Okubo and Levin, 2001). Turchin (1998) recognized that even highly non-random movement of animals can be approximated by diffusion models at some spatial scales. He suggested that diffusion modeling provides a spatial component to mathematical models like the Lotka–Volterra equations. While these equations describe population growth, death, and interaction, diffusion terms redistribute the results in space (Turchin, 1998). Simple diffusion models are limited to random movement in homogeneous space, but the diffusion analogue can also be applied to non-random movements in heterogeneous environments (Reeves and Usher, 1989; Ims, 1995; Turchin, 1998; Okubo and Levin, 2001) (Fig. 11.9). These complex diffusion models link "the behavioral responses to spatial heterogeneity at the individual level, and the spatial distribution of organisms at the population level" (Turchin 1998; p. 49). Okubo and Levin (2001; p. 10) described diffusion as "a phenomenon by which the particle group as a whole spreads according to the irregular motion of each particle." They recognized that animals can be examples of particles in this context.

Reeves and Usher (1989) used a diffusion model to simulate the spread of the coypu (*Myocastor coypus* Molina.) in southeastern Great Britain. The coypu is a rodent that first became established in East Anglia after escaping from

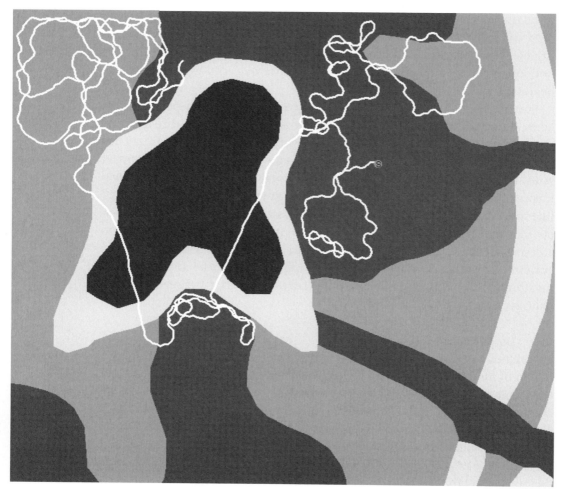

FIGURE 11.9
The concept of a random walk can be used to model animal movement in
heterogeneous environments. In this example, as in the example model for this
chapter, animal movement is simulated by choosing "turning angles" from a
distribution of probable angles. Turning angles vary in different habitat types.

fur farms in the 1930s. It spread rapidly, damaging crops, wild plants, and
drainage systems (Reeves and Usher, 1989). Reeves and Usher (1989) used a
discrete diffusion model based on 10 km grid cells to simulate the spread of the
coypu population. They modified the basic diffusion equations first presented
by Skellam (1951), using methods presented by Dobzhansky *et al.* (1979), to
allow for a discrete (gridded) domain rather than diffusion in continuous space.
Assignment of different carrying capacities and diffusivity parameters to each
grid cell allowed characterization of spatial heterogeneity. In the model, the
coypu population grew according to a logistic population growth routine and

diffused according the discretized diffusion equations (Reeves and Usher, 1989). The model was able to reproduce "a reasonable representation of the spread of coypu," although limitations were noted caused by the coarse grid resolution, changes in diffusivity of cells over time, and the inability of the model to account for human management (specifically, trapping pressure) (Reeves and Usher, 1989).

Individual-based models and random walks

Another diffusion-based method uses computer simulations of paths followed by individual organisms as they move through spatially explicit mosaics (Ims, 1995) (Fig. 11.9). These individual-based, random-walk models are probably the most frequently used way of simulating animal movement in heterogeneous space. Random-walk models operate on discrete time steps during which the "animal" moves for some distance in a single direction. The distance traveled by the simulated animal during each time step can be fixed or variable. In simple random-walk models, the direction of movement is chosen randomly. First-order, correlated random walks, however, account for the tendency of animals to move preferentially in some directions, partly because of their physical characteristics (e.g. bilaterally symmetric animals with heads are more likely to move "forward" than "backward") (DeAngelis and Yeh, 1984).

Simple and correlated random-walk models are typically set in grids or lattices, with cell dimensions corresponding to the environmental grain size recognized by the modeled organism (Ims, 1995). Spatial heterogeneity is represented by, for example, binary schemes that classify some cells as "suitable" and the rest as "unsuitable" (e.g. Turner *et al.*, 1993). Aggregation of similar cells in these grids creates larger structures and has led to the exploration of percolation theory for studying the effect of scale on animal movement (e.g. Gardner *et al.*, 1989). Cells can also be non-binary representations of relevant environmental properties (e.g. vegetation) and given parameters affecting movement. For example, animals may move more slowly through some cover types than others (e.g. Schippers *et al.*, 1996). We describe two examples of random-walk models to demonstrate implicit (Siniff and Jessen, 1969) and explicit (Schippers *et al.*, 1996) treatment of spatial heterogeneity. There are many other examples of this type of modeling in the literature (e.g. Jones, 1977; Bovet and Benhamou, 1988; Crist *et al.*, 1992; Boone and Hunter, 1996), and Turchin (1998) has provided an excellent review with historical notes.

Siniff and Jessen (1969) described the development of a random-walk model for simulating animal movement within home ranges. Their modeling was based on empirical data gathered using field telemetry to track the spatial locations of snowshoe hares and red foxes at specific times. These data enabled the modelers to construct probability distributions describing the distance and direction of travel between observations, and the duration of periods of rest and movement. By limiting the model to a portion of the animals' life spans, Siniff and Jessen were able to ignore movement complicated by migration, breeding, and changes of behavior (resulting, for example, from transition of animals from juveniles to adults). They experimented extensively to find statistical distributions that best described each of the empirical movement parameters. They also experimented with methods for preventing simulated animals from "wandering off" and leaving home ranges. While Siniff and Jessen (1969) were not able to match exactly the movement patterns seen in empirical telemetry data, they were able to produce reasonable approximations. Perhaps more importantly, they made considerable progress towards understanding the effects of various statistical movement distributions and ways of treating home range areas.

The Siniff and Jessen (1969) model did not explicitly simulate hetero-geneous environmental space, although heterogeneity was implicit in the tele-metry data. In contrast, Schippers et al. (1996) developed a GIS-based correlated random-walk model (GRIDWALK) to account explicitly for spatial features, including variability in habitat quality and barriers to movement (i.e. roads and rivers). They used the European badger (Meles meles) as an example animal and simulated movement between metapopulations (setts) in the Netherlands. In their model, 1 km² grid cells contained information about habitat quality (land cover, sett potential) and the presence or absence of badger populations. Linear barriers (roads, rivers, and canals) were located between grid cells and increased the chance of death for badgers moving across them. Badger residence time within each cell was determined by habitat quality, and movement between cells was stochastic but influenced by the direction of previous movement (a correlated random walk). Badger mortality was a combination of background mortality, mortality associated with each grid cell type, and mortality from crossing barriers. Finally, connectivity between metapopulations was calcu-lated by determining the proportion of animals leaving one sett and arriving at another. Schippers et al. (1996) found that the model was "especially sensitive to the response of the animal to landscape quality and barrier parameterization" and they suggested that it is important to quantify those parameters for animal movement modeling.

Not all individual-based models use random walks to simulate animal move-ment. Turner et al. (1993) simulated foraging behavior of groups of ungulates in Yellowstone National Park using a gridded landscape (1 ha cells) and rule-based animal movement. Rather than moving according to a probability distri-bution as in random-walk models, ungulates in this model moved in response to their perception of the spatial distribution of resources. Cells were classi-fied as either resource sites or non-resource sites, with resource sites containing

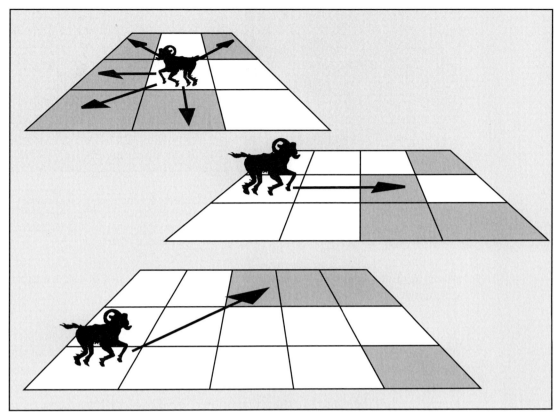

FIGURE 11.10
Movement rules used in the Turner *et al.* (1993) model. In this cell-based scheme, animals could move only to adjacent cells (top), to the closest resource-containing cell within a pre-assigned search radius (middle), or in the direction that contains the most resource sites within the search radius (bottom), even if some other direction offers a closer resource site.

a finite amount of food that could be reduced or removed by browsing. Three movement rules were tested in the model (Fig. 11.10). In the first, the animals could move only to an adjacent cell at each time step. They chose which direction to move by evaluating the number of resource sites in four wedge-shaped areas radiating outwards from the animal and with a radius set according to the theoretical "resource utilization scale" of the modeled animal. Animals moved one cell in the direction of the most resource-rich wedge. The other rules allowed variable movement lengths up to some maximum per time step. In one, called the closest-resource-site rule, the animals move to the closest resource-containing cell within its search radius. In the other, called the best-direction rule, the animals move in the direction of the greatest number of resource sites, even if some other direction offered closer resources. The

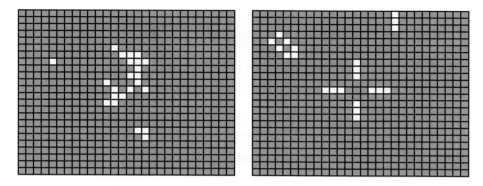

FIGURE 11.11
The "game of life," a simple cellular automata scheme for simulating the spread of
populations based on simple rules. Cells initially exhibit two states, "on" or "off."
A cell initially "on" stays "on" if two or three of its neighbors are "on". Cells that are
initially "off" can be switched "on" if three of neighbors are "on."

model included detailed routines for calculating energetics of the foraging
ungulates.

Cellular automata

Another way of modeling animal movement, particularly in the context of the
spread of populations, is to use cellular automata. In these models, the state
of any cell is determined only by the state of cells in its immediate neighborhood. Turchin (1998) cited "the game of life," developed by John Conway, as
an example of a very simple population spread model that uses the cellular
automata approach (Fig. 11.11). In this model, cells can exhibit two states, "on"
or "off." A cell that is initially "on" stays "on" if two or three of neighbors are
"on". Cells that are initially "off" can be switched "on" if three of neighbors are
"on". Simple models like this can produce surprisingly rich behaviors depending on initial conditions. According to Turchin (1998), four types of long-term
dynamic can be generated including (i) homogeneity with the disappearance of
the initial pattern, (ii) periodicity, (iii) chaos, and (iv) complex regional patterns
and localized structures that can move across the domain. One could imagine
using more complex rules or simulating specific motivating forces (e.g. forage
quality, taxis) in a cellular automaton to explore different types of transport. In
fact, many diffusion models are based on this approach.

11.4 Introduction to the example model

The American (pine) marten (*Martes americana*) is found from Newfound-
land and Nova Scotia, west to Alaska, and south into the Rocky Mountains

and California. They are also found in parts of Maine, New York, Michigan, Minnesota, and Wisconsin (Nowak, 1991). Martens are members of the weasel family (Mustelidae) and are typically about the size of a house cat, with gray heads, dark brown or black legs and tail, light brown backs, and a cream-colored patch on their chests. Mature forests are considered ideal marten habitat (Buskirk and Powell, 1994) because of decreased risk of predation, increased abundance of subnivean (below the snow) resting sites, and increased prey availability relative to less-mature forests (Chapin *et al.*, 1997).

Forest fragmentation, particularly from clearcutting, has been identified as a potential threat to American martens, because risk of predation increases when martens cross regenerating clearcuts. Additionally, small remnant patches of old growth forest in clearcut landscapes may not be large enough to provide adequate resources to balance the risks for martens accessing them (Chapin *et al.*, 1997). These authors suggested that "quantitative relationships between landscape pattern and spatial use of habitat are needed to evaluate whether forest carnivores perceive industrial forest landscapes as fragmented at scales such as the forest patch and the individual home range, and to improve landscape-scale management of forested habitats." Simulation models of marten habitat use are tools for examining these relationships.

A spatially explicit model of American marten movement through areas of different habitat quality was created based on data collected by S. W. Buskirk and G. Beauvais (unpublished data) in the Medicine Bow Mountains of Wyoming. They measured turning angles of pine martens at intervals of 5 m as the martens moved about in various cover types. Cover was classified into four habitat quality classes (high, low, open, water) based on old growth characteristics (or lack thereof). For each category, probability distributions were created using mean turning angle and standard deviation.

The model is based on the correlated random-walk approach. A simulated marten is given a starting point on a tessellated (raster) map depicting the spatial arrangement of the four habitat classes. The model reads the habitat class from this raster GIS layer and chooses a turning angle from a Gaussian probability distribution for that class. The marten moves in the indicated direction for a fixed distance. The process is repeated for a given number of time steps, and the resulting path is displayed on the habitat map in the GIS.

To run the model, the user first chooses one of three landscapes to serve as a domain for marten movement from a browse window. The first depicts a clearcut landscape dissected by a road and a river. Clearcuts vary in size, shape, and pattern. The second landscape choice is more "natural" and is derived from actual land-cover data from the Medicine Bow National Forest in Wyoming. Forested and open terrain occur along with a lake, and habitat units are relatively large.

The third landscape includes a divided highway with "crosswalks" of high-suitability habitat, a strategy that has been implemented in Banff National Park in Canada. The model used can examine the effect of any one of these landscape pattern scenarios on martens moving in response to the "rules" defined by turning angle probabilities.

After choosing a hypothetical landscape, the user decides whether to employ turning angle distributions based on the empirical data (S. W. Buskirk and G. Beauvais, unpublished data) or on randomly generated turning angles. By comparing output from these two scenarios, the user can explore whether habitat quality influences the amount of time martens spend in a particular habitat type. Users also have the option of changing the probabilities of martens moving from one habitat type to another. These probabilities adjust the impedance between cover types in the landscape by increasing or decreasing the probability of martens turning when they encounter a boundary.

The user chooses a starting point for marten movement in the landscape by pointing with the mouse. Finally, the number of time steps and the distance traveled by the marten at each time step are chosen and the model is started. The model runs, and the simulated path taken by the marten through the landscape is displayed in the interface. The user can query the interface using standard ArcView tools to explore the effects of the various scenarios on the outcome.

Models are *depictions* of reality constructed by reducing real-world processes to a simplified set of governing rules or equations that can be expressed in mathematical or statistical terms. This usually requires simplifying assumptions, violation of which can cause divergence of the model results from "truth." In the model described here, habitat patches are assumed to be internally homogeneous and to have discrete boundaries. Water is not treated as a barrier to American marten movement, even when water "patches" are large. There is no functionality in the model for describing the effects of traffic on the roadways on marten movement. Predation of martens and chance encounters with other animals are not considered. Are these assumptions important? This, of course, depends on the questions the modeler is asking and the temporal and spatial scales at which these questions operate.

11.5 Summary

Animals move through space across a wide range of scales for a variety of reasons and in so doing affect almost every aspect of ecosystem structure and function. Animal movement is unique among the transport vectors discussed in this book because, by its very nature, it cannot be modeled deterministically.

As a consequence, models of animal movement are based at least partially on stochastic elements and probabilities. In this chapter, we introduced some of the contexts in which animal movement has been studied, and a few of the important approaches that have been used to model their movement. We also provided a simple model of pine marten movement based on random-walk algorithms.

Electromagnetic radiation

The sun rises. In that short phrase, in a single fact, is enough information to
keep biology, physics, and philosophy busy for the rest of time.

Lyall Watson (1982)

On a warm summer night, an adult female firefly, perched on a blade of grass,
watches as airborne males pass overhead. Attracted to a series of light flashes
characterizing a male of her species, the female responds in kind. A brief
exchange of confirming flashes ensues, and the male approaches the female,
using her light as a locator beacon (McDermott, 1958; Lloyd, 1966). Other
species of fireflies produce bioluminescence only as larvae, warning potential
predators of their unappealing taste (caused by defensive chemicals). The predators do not approach, instead searching for other, more palatable food (Sivinski,
1981).

Early morning light changes in color and intensity as the sun rises in the
sky and is filtered through different combinations of leaves and forest gaps
before illuminating the forest floor. Clouds occasionally intercept the direct
rays of sunlight, further altering the light quality (Endler, 1993). A red Indian
paintbrush (*Castilleja coccinea*) blossom growing in the understory is highlighted
for a few minutes by a "sunfleck," attracting the attention of a ruby-throated
hummingbird (*Archilchus colubris*). The hummingbird feeds on nectar from the
flower, inadvertently collecting pollen grains, which it later transfers to another
flower elsewhere in the forest.

These interactions are examples of light originating or changing quality
at one location in space and causing an effect at another. In this chapter, we
discuss transport by light. In particular, we emphasize light as a signal carrying
information, which, upon translation, affects the behavior of the organism
receiving it.

12.1 Transport system description

Electromagnetic radiation (EMR), including visible light, travels
through air and water and reflects off solid objects in the landscape. It carries

with it information that can be received and interpreted by animals that have evolved organs specialized for this purpose – eyes. A common example of the transport of information by EMR begins when light from the sun reaches the earth, reflects off an object, such as a patch of blackberries, and is received by the eyes of an animal, such as a hungry bear. As a result of this transport event, the bear finds her breakfast. Animals use the information carried by visible light to find food and shelter, to avoid predators, and to communicate with each other. They produce visual signals to advertise their presence, to declare territorial boundaries, to indicate aggression, and to attract mates. Plants also use reflected light to attract animals for pollinating their flowers and dispersing their seeds. Thus, the transport of information by EMR has varied and important ecological effects.

12.2   Underlying principles for mode of transport

12.2.1 Physical principles of electromagnetic radiation and light

EMR is radiant energy that moves in transverse waves, generating electrical and magnetic forces as it passes a given location. It is produced when the motion of electrons within atoms and molecules is accelerated by an external source of energy. Vibrating electrons generate an electric field, which, in turn, induces a magnetic field, and the coupled electric and magnetic oscillations move outward from the source (Bradbury and Vehrencamp, 1998) (Fig. 12.1). Electromagnetic waves can be described by their frequency (the number of waves passing a point in a given time interval), which is inversely proportional to wavelength. Naturally occurring EMR ranges in wavelength from the longest radiowave (≥ 1000 km), through microwaves, infrared, visible light, ultraviolet, X-rays, gamma rays, and to the shortest cosmic rays (≤ 0.001 nm) (Fig. 12.2). Visible light occupies the intermediate wavelengths around 300–800 nm (Bradbury and Vehrencamp, 1998). The sun is the most obvious source of EMR in the natural world, providing heat to warm the earth and light needed for photosynthesis.

In some ways, EMR behaves like sound (Ch. 13): its intensity decreases as the inverse of the squared distance from its source, and attenuation with distance depends on frequency and the medium through which it travels. EMR travels at different speeds in different media and, as a result, exhibits the wave properties of reflection, refraction, and diffraction. In other respects, EMR behaves like a stream of particles. These particles can be viewed as packets of energy, called quanta, with the energy in each a function of the frequency of the EMR. The higher the frequency (and thus the shorter the wavelength), the higher the energy per quantum. Unlike sound, EMR is not a pressure wave; it

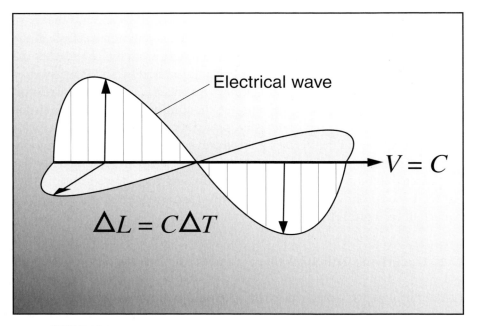

FIGURE 12.1
An electromagnetic wave. Light comprises orthogonal electrical and magnetic waves.

requires no medium for transmission and actually travels fastest in a vacuum (at approximately 3×10^8 m/s) (Bradbury and Vehrencamp, 1998). Visible light travels in a straight line through air and water, reflects off most objects in the environment and then continues to travel in a straight line but in a new direction. Because of these properties, visual communication between animals is blocked by opaque objects directly between them.

12.2.2 Production and transmission of light signals

When visible light is present, plants and animals can passively generate visual images as light reflects from them to the eyes of other animals (Fig. 12.3). Animals can make themselves more or less conspicuous by increasing or decreasing the contrast between themselves and their background using color, pattern, and movement (Endler and Thery, 1996). For example, the common combination of yellow and blue in brightly colored reef fishes may be useful for communication (e.g. territorial signaling) if displayed against a monochromatic background, but it is also useful for camouflage if the fishes position themselves among branched coral (Marshall, 2000). Some cephalopods, fish, amphibians, and reptiles can change their body color or pattern for camouflage or

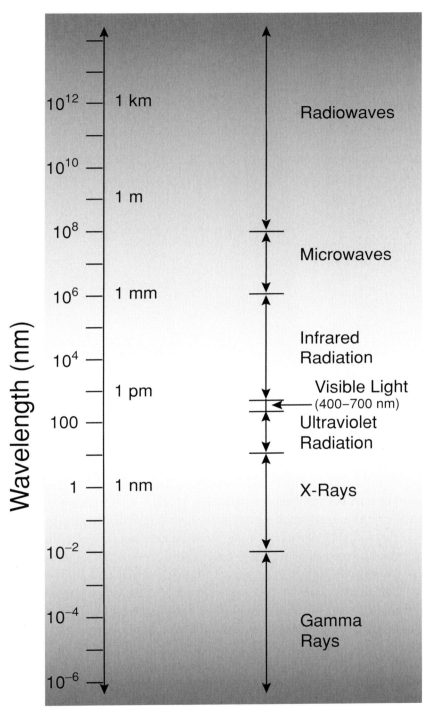

FIGURE 12.2
The electromagnetic spectrum.

FIGURE 12.3
Zebra coloration helps them to avoid predation by providing camouflage. Each
zebra has a unique pattern of stripes.

social signaling. Other animals hide or reveal flashes of color on their body
with varying postures or choose a particular spectral environment that will
enhance the visual signals they wish to send. Lekking birds from the forests of
French Guyana choose particular light environments to maximize the visual
impact of their colorful displays (Endler and Thery, 1996). Males of two species,
Rupicola rupicola and *Corapipo gutturalis*, display only when a patch of sunlight
is present on their chosen lek site, while males of *Lepidothrix serena* mainly
display when the sun is below the horizon or obstructed by clouds (Endler
and Thery, 1996). The color and patterns of many flowers have evolved to be
attractive to their respective pollinators. Insect pollinators such as bees can
discriminate colors and this probably contributes to flower constancy, a phenomenon in which the insects restrict their visits to one flower type even in
the presence of other rewarding flowers (Dyer, 1998). Flower constancy benefits the flowering plant by increasing the chance of cross-fertilization (Waser,
1986).

A diverse range of organisms (e.g. bacteria, fungi, glow worms, beetles, fire-flies, cephalopods, jellyfish, fish) can generate their own light via biolumines-cence, a process in which molecules of luciferin are excited when electrons are raised to a higher orbital by the energy-transporting chemical adenosine trisphosphate (ATP). When the electrons fall back to their baseline position, photons are emitted and light is produced. This form of communication is mostly used by nocturnal organisms or those that live in dim environments such as deep water (Bradbury and Vehrencamp, 1998). In the oceans, light becomes dimmer and bluer with increasing water depth, as water strongly absorbs both ultraviolet and infrared radiation; below 1000 m, almost no daylight remains. At these depths, the only visible light is from point-source bioluminescent sig-nals produced by other animals (Warrant, 2000). These signals typically consist of flashes of varying color, intensity, and frequency. Deep-sea animals, including lanternfish, anglerfish, and cephalopods, express sexual dimorphism of photo-phores (light-producing organs) and signaling patterns to facilitate sexual encounters (Herring, 2000). For example, females of the anglerfish *Chaenophryne draco* have an elaborate luminous anterior lure, while males have no lure (Herring, 2000).

Difficulties associated with long-distance transmission of visual signals are similar to those encountered with long-distance sound transmission: attenu-ation, pattern loss, and background noise. Transmission of light signals also depends on the availability and quality of ambient light, and the absence of opaque obstacles between the sender and receiver. Light from the sun is scat-tered, filtered, and attenuated as a function of wavelength, sun angle, atmo-spheric properties, weather conditions, and surface properties. This results in a light spectrum that is reduced in some spectral regions (Fig. 12.4), and senders can only reflect those wavelengths that are available. The spectrum of light reflected from the sender depends on the characteristics of its surface, while the conspicuousness of the signal is a function of the perceived contrast between the light reflecting from the sender and that reflecting from the background (Bradbury and Vehrencamp, 1998).

The combined effects on a light signal of global attenuation, medium absorp-tion, filtering, and scattering can be represented by a single parameter: the beam attenuation coefficient (Bradbury and Vehrencamp, 1998). The recipro-cal of this coefficient is attenuation length, the maximum distance at which a large and contrasting object can just be detected. Attenuation length depends on the wavelength of the signal and the medium through which it passes. In air, attenuation length decreases with wavelength. For example, a light signal with a wavelength of 600 nm (yellow–orange) has an attenuation length of 120 km in pure air, while that of a 500 nm (blue–green) signal is only 55 km (Bradbury and Vehrencamp, 1998; taken from Dusenbery, 1992).

FIGURE 12.4
Light attenuation across heterogeneous space.

12.2.3 Principles of vision

To use EMR for communication, organisms must detect it. This is a two-stage process; the electromagnetic energy must first be absorbed and trapped by a receptor molecule, and then the resulting changes in the molecule must be used to stimulate an electrical response in a receptor neuron. Only a narrow range of the available wavelengths of EMR is suitable for vision (Fig. 12.2). Radiowaves have such low frequencies and low energy that they pass through or around most biological entities without being absorbed. Microwave radiation tends to be absorbed by water molecules in the atmosphere; the levels of ambient radiation at these frequencies are too low for them to be useful for biological vision. Infrared radiation is perceived by animals as heat. Many types of molecule can absorb infrared and become warmer. In addition, all animals (indeed all things) in an environment warmer than absolute zero produce their own infrared radiation as a result of vibration of atoms within their molecules. Therefore, a visual system based on infrared is possible, although difficulties include the quick absorption of infrared by biological tissues, which reduces the amount of energy reaching receptor molecules, and the complexity and variability of the thermal world. Some snakes, such as pit vipers, puff adders and boas, have evolved infrared sensors in their nasal region for detecting warm-blooded prey (Newman and Hartline, 1982; York *et al.*, 1998). Such heat detectors work only at relatively close range and have limited resolution and sensitivity (Bradbury and Vehrencamp, 1998).

Radiation in the visible light range contains sufficient energy to shift electrons within the atoms of a receptor molecule to a higher energy level and thus produce the configurational changes that can be coupled to the excitation of nerves. However, unlike ultraviolet radiation and the shorter wavelengths of EMR, visible light does not contain sufficient energy to break chemical bonds and damage biological molecules. Objects are visible when they reflect EMR. Most solid objects reflect visible light while transmitting both longer and shorter electromagnetic waves. Furthermore, most of the EMR reaching the earth from the sun falls within the visible light range; consequently, the frequency range most useful for biological vision is also that most readily available (Bradbury and Vehrencamp, 1998).

Animal eyes use visible light to form an accurate image of the shape, size, color, and proximity of objects in their environment. Visible light enters the eye and is refracted by a lens to form a spatial image of the location of all reflecting objects in the landscape (Fig. 12.5). Eyes have evolved to suit the different habitats, diets, and activity patterns of different animal species. For example, nocturnal animals must maximize the sensitivity of their eyes to low light levels and, as a consequence, tend to have reduced spatial and temporal resolution

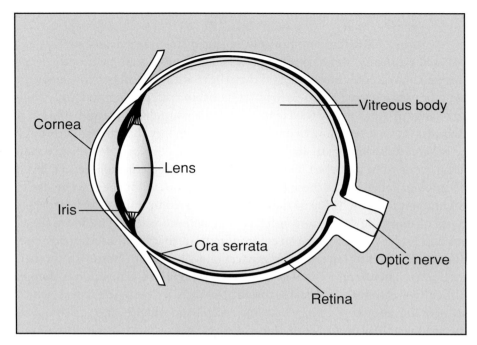

FIGURE 12.5
The human eye.

of visual images (Bradbury and Vehrencamp, 1998). Zollner and Lima (1999) demonstrated that white-footed mice released in an open field on a dark night could not perceive patches of forest 30 m away. However, the perceptual range of the mice increased to approximately 60 m in full moonlight. Predators require good depth perception and spatial and temporal resolution so that they can detect the movement of a prey animal and accurately judge the distance between it and themselves (Bradbury and Vehrencamp, 1998).

In the darkness of the deep sea, fish require sensitive eyes that can accurately locate the bioluminescent flashes of potential prey or suitable mates in order to ensure successful predation and sexual encounters (Warrant, 2000). Fiddler crabs (*Uca* spp.) live in burrows on sand and mud flats and have compound eyes on long, vertical stalks (Fig. 12.6). They use vision to detect predators and conspecifics, with males possessing one hugely enlarged claw for social signaling. The eyes of fiddler crabs have evolved to suit the flat world in which they live, with vertical resolution maximized in the middle or equatorial region of their visual space, where most relevant events are likely to occur (Zeil and Zanker, 1997). Although humans cannot perceive ultraviolet light (wavelength < 400 nm), a wide range of other animals, including insects, crustaceans, birds, and some mammals, possess ultraviolet receptors within their visual systems,

FIGURE 12.6
Fiddler crab (*Uca vocans*) eyes are located at the ends of stalks, which serve to increase the viewshed visible at any time to the crabs. This allows crabs to see predators earlier than would otherwise be possible.

using them for color discrimination, navigation, and calibration of circadian rhythms (Tovee, 1995).

12.3 Modeling techniques

There is a paucity of models explicitly simulating the transport of information or energy by light in ecological settings. While many models predict the amount of solar radiation arriving at a particular place (e.g. Bristow and Campbell, 1984; Thornton and Running, 1999) or the behavior of light interacting with objects (e.g. reflectance) (Kowalik *et al*., 1982; Pech and Davis, 1987), we are more concerned here with light transport. This type of transport primarily involves signaling – either among animals or between plants and animals – and is discussed in the animal communication literature (e.g. Bradbury and Vehrencamp, 1998) or in descriptive literature treating, for example, plant–pollinator relationships (e.g. Nilsson, 1988). In these contexts, models are usually conceptual and they rarely include explicit depiction of spatial relationships. In this section, we give examples of a few approaches that do treat light as a transport vector in spatial contexts or that have potential for modeling applications.

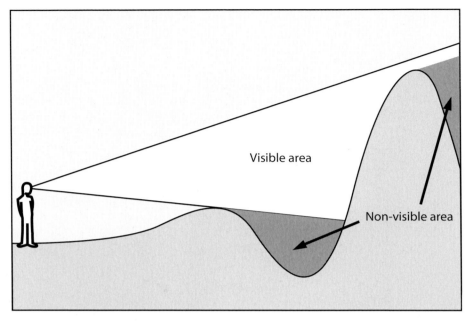

FIGURE 12.7
A viewshed is the area that is visible by a viewer at a given point in space. The extent
of viewsheds is determined by factors including terrain, vegetation, and viewer
height.

12.3.1 Viewshed analysis

Visual communication between animals can be modeled across three-dimensional landscapes using GIS tools such as viewsheds and horizons (e.g. Germino *et al.*, 2001). A viewshed represents the parts of a terrain visible from a particular point in the landscape, while a horizon describes the distal boundary of the viewshed (De Floriani and Magillo, 1999) (Fig. 12.7). Viewshed analysis can be used to determine whether an animal located at one place on a terrain is visible from another, or to predict the area within an animal's visual range that might include, for example, potential prey (or predators). Viewshed analysis is complicated by the presence of vegetation, which obscures views at scales finer than the resolution of most spatial data available to modelers. This is especially true in forests but might be less of an issue in, for example, African grasslands, where many predator and prey species are taller than the dominant vegetation (Fig. 12.8). Even in the latter situation, predators use patches of tall vegetation to disguise their presence, essentially removing themselves from the viewshed of their prey. One could imagine models of view characteristics in spatial contexts that could predict, for example, preferred migration corridors for grassland ungulates moving through woodlands with sparse trees. Such models would require highly resolved vegetation maps to be realistic.

FIGURE 12.8
Cheetahs (*Acinonyx jubatus*) in Masai Mara National Park, Kenya increase their viewshed by resting atop a well-defined mound. Their position on this mound affords them much more extensive views than if they were to rest among the tall grass that covers the majority of the landscape here.

12.3.2 Light quality

The quality of light at a particular place varies with the time of day, weather conditions (particularly cloudiness), and the arrangement of objects between the sky and the ground (e.g. forest trees, water) (Federer and Tanner, 1966; Messier and Bellefleur, 1988; Endler and Thery, 1996). These variations in light environments in forests were the subject of studies by Endler (1993) aimed at understanding light environments and their potential effects on communication among animals, between animals and plants, on photosynthesis, and on plant morphogenesis. Endler (1993) classified forest light into four groups – forest shade, woodland shade, small gaps, and large gaps – and noted that early or late daylight and cloudiness further modified the light environment. These environments have effects ranging from changes in the ability of animals to remain camouflaged to changes in plant growth and form. Endler and Thery (1996), in a study of forest bird lekking behavior, noted that birds using leks for mating displays may be disproportionately affected by changes in forest structure because of the associated changes in light environments. While Endler does not explicitly model the spatial distribution of light on the forest floor, his work suggests that such models might be instructive (Endler, 1993), a possibility that we have explored with the example model for this chapter.

12.4   Introduction to the example model

A simple forest-floor light-quality model was developed by the authors as the example model for this chapter. The model is based on a three-layer forest canopy through which sunlight passes before encountering the forest floor below. Light quality at the forest floor depends entirely on the number of leaves through which light passes, which is, in turn, determined by sun angle and azimuth, the percentage closure (leaf density) of each of the three canopy layers, and the distance between canopy layers (canopy separation). Model users can adjust all of these parameters to explore their effects on the pattern. As a result, sunlight can pass through either none, one, two, or three leaf layers to create a spatially explicit pattern of differing light quality on the forest floor.

Each canopy layer in the model is populated randomly with leaves until the user-supplied leaf density is achieved. Canopy layers are represented as binary raster layers, with a value of 1 indicating presence of leaves and 0 indicating their absence. A Fortran77 routine uses the sun angle and azimuth to calculate which cells in the middle and lower canopy layers correspond to a sunray passing through each cell in the top canopy layer. For each raster cell in the top canopy layer, the model calculates how many leaves a sunray of given angle and azimuth must pass through to reach the forest floor. ArcView displays this result as a colored pattern.

While crude, the model suggests a basis for experimentation using the spatial analysis tools available in ArcView. What is the safest path across the forest floor for a small mammal attempting to avoid all "sun flecks," where it might be more visible to an airborne predator? What proportion of the forest floor is illuminated by light altered by transmittance through three leaf layers at a given time of the day or year? What are the effects of changes in canopy separation or leaf density on the overall pattern of light on a forest floor? With some minor alterations in the Avenue scripts, one could explore specific canopy characteristics as well. What is the effect of changing the leaf distribution in each layer from random to patterned?

12.5   Summary

Transport by light in ecological systems includes information transmission (signaling) and energy transport (e.g. photosynthetically active radiation).

Despite the importance of light as a transport medium, explicit spatial modeling of light in these contexts is uncommon but may provide new ways of looking at some ecological interactions such as predator or prey viewsheds, the spatial pattern of light quality, and spatial relationships between signaling animals (e.g. Endler and Thery, 1996). Explicit consideration of these spatial relationships may stimulate new ways of exploring ecological pattern.

The propagation of sound

The more elaborate our means of communication, the less we communicate.

Joseph Priestley

13.1 Transport system description

As darkness deepens, Brazilian free-tailed bats (*Tadarida brasiliensis*) scan the night sky for insect prey, not with their eyes but by using echolocation, an active form of "remote sensing." Sounds produced by the bats bounce off of the wings of flying insects and return to the bats' sensitive ears (Fig. 13.1). By perceiving subtle characteristics of the returning sound waves, the bats determine the location, size, and shape of the insects and, in some cases, whether or not the insect is flying towards or away from them (Wilson, 1997). Not all potential bat prey are hapless victims of echolocation, however. Some moths (Lepidoptera) have turned the tables, evolving the ability to sense the sounds made by hunting bats and to determine the *bat's* location and probable path. Such moths make split-second decisions about whether to adopt an evasive flight path or, if time is short, to dive for the ground to avoid capture (Bradbury and Vehrencamp, 1998).

Sound is used by organisms in nature almost exclusively as a vector for information transport. Echolocation by bats is one example. More common are the calls of birds, whales, elephants, and other animals delimiting territory or attracting mates, and the sounds perceived by predators as potential prey move through environmental space. This chapter reviews the concept of sound as a transport vector.

13.2 Underlying principles for mode of transport

13.2.1 What is sound?

Sound is mechanical energy transmitted through a material medium (e.g. air, water, or soil) in longitudinal or transverse pressure waves (Fig. 13.2).

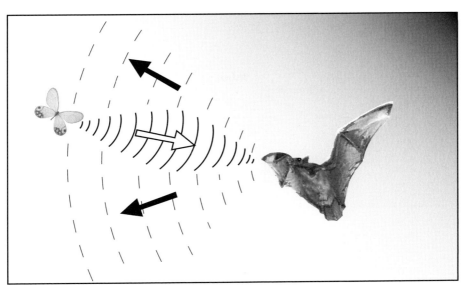

FIGURE 13.1
Bats use a type of sonar to detect potential prey. By sending out sound waves and detecting the characteristics of the reflected sound, bats can locate prey very accurately.

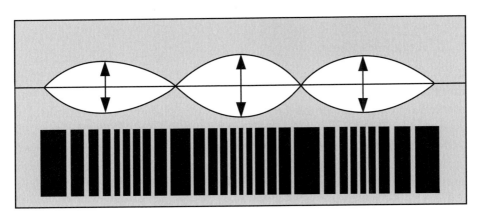

FIGURE 13.2
Sound waves have both transverse and longitudinal characteristics.

Longitudinal waves move molecules back and forth along the axis of wave propagation, while transverse waves move them perpendicular to the direction of wave propagation. Sound is transmitted through gases and liquids in longitudinal waves, and through solids in either longitudinal or transverse waves (Bradbury and Vehrencamp, 1998). Sound is produced by generation of a local concentration or rarefaction of molecules in a medium, which can be recorded as a variation in pressure. This disturbance in local molecular concentration is

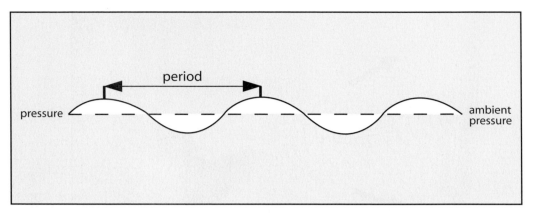

FIGURE 13.3
The form of a sinusoidal pressure wave. Local pressure increases and decreases as the
wave passes a particular point in space.

propagated away from the source of the sound through collisions of successive
layers of molecules; in the absence of barriers, propagation occurs in a sphere of
increasing radius. At each molecular collision, some of the energy in the disturbance is lost as heat. At a sufficient distance from the source, all of the energy will
be dissipated and no sound will be detected (Bradbury and Vehrencamp, 1998).

13.2.2 Properties of sound

The amplitude of a sound can be measured in pressure (pascals (Pa)) or
intensity (watts per square meter); pressure is proportional to the square root
of intensity. As a sound propagates, its energy is spread over a greater sphere of
disturbance. The inverse square law predicts that the intensity of a sound will
decrease as the square of the distance from the source, while sound pressure will
decrease as the reciprocal of distance. This is known as spherical spreading. Both
pressure and intensity can be expressed in decibels (dB), which are logarithmic
units. The decibel scale has been designed to give the same values for sound
pressures and intensities. Spherical spreading leads to a decrease of 6 dB in
both the pressure and the intensity of a sound for each doubling of distance
from the source (Forrest, 1994; Bradbury and Vehrencamp, 1998).

The simplest sound is one in which pressure varies sinusoidally (as a
sinewave) over space and time (Fig. 13.3). The temporal pattern of rise and
fall in signal pressure is called the waveform of the signal, the time interval
between successive sinusoidal peaks (the duration of the sound wave) is known
as the period, and the number of cycles per unit time is the frequency of the
wave. Frequency (in cycles/second) is measured in hertz (Hz). The wavelength
of a sound is the physical distance between successive peaks of a sound wave.

FIGURE 13.4
African elephant in Samburu National Park, Kenya. Elephants communicate using
very-low-frequency sounds.

For a given medium, frequency and wavelength are inversely related: high
frequencies have short wavelengths and low frequencies have long wavelengths.
For example, the call made by male African elephants during musth, a period of
heightened aggression and sexual activity, has a fundamental (lowest-pitched)
frequency of 12–17 Hz (Poole, 1999) (Fig. 13.4). In the atmosphere, the wavelength of this sound is 20–29 m at 20 °C. In contrast, the call of the Australian
field cricket, *Teleogryllus oceanicus*, has a frequency of 4.5 kHz (Hill *et al.*, 1972),
which corresponds to a wavelength of approximately 0.076 m (7.6 cm) in the
atmosphere.

13.2.3 Animal sounds

The natural environment is full of sounds of abiotic and biotic origin,
such as those generated by wind, rain, thunder, flowing water, and calling animals. Insects, fish, frogs, birds, and mammals produce a wide range of acoustic
signals, which serve a diversity of biological and social purposes. These include
attracting and bonding with mates, protecting territories, maintaining spacing

between groups, communicating distress and hunger, and warning of danger from predators (e.g. MacKinnon, 1974; Wasser, 1977; Duellman and Trueb, 1986; Strahan, 1996; Poole, 1999).

Production of an acoustical signal can be energetically costly (Prestwich, 1994) and will only benefit the sender if the signal is successfully transmitted to the intended receiver or receivers (Brenowitz, 1986). However, the size of an animal and the environment in which it calls place considerable constraints on acoustic communication, particularly over long distances (Bradbury and Vehrencamp, 1998). The distance over which an acoustic signal is effective depends on the characteristics of the signal, the environment through which it travels, the level of ambient noise, and the ability of the receiver to extract information from the transmitted sound (Forrest, 1994). Propagation of sound in terrestrial habitats is influenced by the amplitude, frequency, and complexity of the sound; by vegetation, topography, and weather conditions; and by the proximity of the sender and receiver to a boundary such as the ground or a forest canopy (Larom *et al.*, 1997a,b; Bradbury and Vehrencamp, 1998).

13.2.4 Propagation of animal sounds

As outlined above, the amplitude of a sound decreases or attenuates as a function of distance from the source. In natural habitats, sound generally attenuates at a rate in excess of that predicted by spherical spreading alone. This excess attenuation can be caused by atmospheric absorption, scattering, or boundary interference. Sound energy is absorbed by the atmosphere through heat conduction, viscous losses, and molecular absorption. The amount of sound energy lost in this way increases with the frequency of the sound, but it also varies with the humidity of the air. For frequencies below 8 kHz, attenuation increases approximately linearly as humidity decreases (Wiley and Richards, 1978). Vegetation, topography, and atmospheric turbulence can scatter sound waves, deflecting sound energy from the path of propagation and causing interference between the direct wave and the scattered waves. Wavelengths shorter than or equal to the dimensions of environmental obstacles (e.g. leaves, tree trunks, boulders) will be scattered more than longer wavelengths. Consequently, high frequencies will attenuate more rapidly as a result of scattering than will low frequencies (Brenowitz, 1986). Ambient noise produced by wind, turbulence, and the calls of other animals can interfere with the detection of a signal and transmission of its information. Signal detection and discrimination are better at higher signal-to-noise ratios (Forrest, 1994). The movement of wind along the ground, through vegetation, or past the head of an animal generates predominantly low-frequency noise, with most energy contained below 2 kHz (Brenowitz, 1986). Therefore, for a given level of signal energy, the signal-to-noise ratio will be higher at frequencies above 2 kHz (Forrest, 1994).

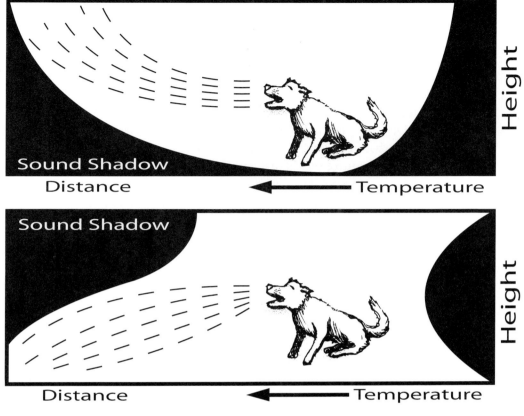

FIGURE 13.5
Sound can be concentrated into sound channels or directed away from sound shadows by temperature gradients.

Propagation of sound at boundaries such as the surface of the ground or a lake depends on the acoustic impedances of the adjoining media, the spatial arrangement of the sender and receiver, and the wavelength of the sound (Fig. 13.5). Acoustic impedance is the product of the density of the propagating medium and the speed of sound in that medium; it is expressed in rayls (1 rayl = 1 N s/m^3). When sound traveling in a medium (e.g. air) with one acoustic impedance, encounters the boundary of another medium (e.g. soil, rock, or water) with a different acoustic impedance, some of the incident energy will be reflected back into the first medium and some will penetrate into the second medium. If the latter passes from a medium of low velocity such as air to one of higher velocity such as soil, the sound wave will be bent (refracted) towards the boundary; if travelling from a medium of high velocity to one of lower velocity, the sound wave will be refracted into the second medium (Bradbury and Vehrencamp, 1998).

The greater the difference between the acoustic impedances of the adjoining media, the greater the proportion of sound that will be reflected at the boundary, and the smaller the proportion that will enter the second medium (Bradbury and Vehrencamp, 1998). For example, air has an acoustic impedance of 30 rayl. If sound propagating through the atmosphere encounters a boundary of grass-covered ground with an estimated impedance of 150–300 rayl (Embleton *et al.*, 1983), less sound will be reflected, more will penetrate into the ground, and attenuation will be higher than if that sound encountered a boundary of water, with an acoustic impedance of 150 000 rayl (Bradbury and Vehrencamp, 1998). When both the source of a signal and the receiver are on the ground, excess attenuation over soft surfaces with low acoustic impedances can be very large, and the effective range of the signal will be reduced. Conversely, if the sender and receiver are on the ground but the signal travels over water, signal amplitude can be increased by 6 dB above that expected from spherical spreading, doubling the distance of propagation (Forrest, 1994). When the source or the receiver are elevated above the ground, the attenuating effect of a soft ground surface is reduced (Marten and Marler, 1977; Forrest, 1994). Thus, calling from an elevated perch can dramatically increase the effective distance of a signal (e.g. male short-tailed crickets calling on elevated perches had an effective signal area 14 times that of crickets calling on the ground (Paul and Walker, 1979).

As outlined above, sound energy that crosses a boundary between two media of different acoustic impedances is likely to change its direction of travel. Gradients of temperature, pressure, composition, and mass flow levels are often found in air and water, and these also generate gradients of sound velocities. Consider a gradient as a series of boundaries between layers of different sound velocities. As a propagating sound moves through each layer, it is bent successively in the same direction, producing a curved trajectory instead of a straight one. If it moves from an area of low to one of high sound velocity, the direction of propagation will be bent back towards the layer in which it began; if it moves from an area of high to one of low sound velocity, it will be bent away into the second area. Increasing temperature or pressure increases the speed of sound. Sound propagating with the wind has a higher velocity than that propagating against the wind (Bradbury and Vehrencamp, 1998).

On a sunny day, the air nearest the ground is warmed by the surface of the earth, and air temperature decreases with increasing height (Fig. 13.5). The call of an animal sitting on the ground will be refracted up into cooler air where sound velocity is lower, producing a sound shadow near the ground in which the call cannot be heard. Conversely, on a clear, still night, the surface of the earth radiates heat into the atmosphere, forming a temperature inversion in which the air nearest the ground is cooler than the air above it. The call of an animal within the cool layer will radiate up into the warmer air and then be refracted back towards the earth. This produces a sound channel, which can trap sounds and propagate them over long distances (Fig. 13.5) (Bradbury and Vehrencamp, 1998). For example, the range over which African elephants can communicate in the savannas of Namibia more than doubles at night, and it is optimized on clear, calm, and relatively cold nights when strong temperature inversions form (Larom *et al.*, 1997b) (Fig. 13.4). Under these conditions, their low-frequency calls (around 14–35 Hz) may be detected by other elephants more than 10 km away. Remarkably, under certain conditions, elephants may be able to hear the rumble of thunderstorms 250 km away and respond by moving towards them before other animals sense the potential onset of rains. Dawn and evening calling in other species, such as lions, wolves, frogs, insects, and birds, may also be explained by enhanced sound transmission at these times (Larom *et al.*, 1997a,b). In a forest, the warming of air during the morning establishes a temperature gradient that influences sound propagation. In the first few hours after sunrise, air at the top of the canopy and above it heats up, while air below the canopy remains relatively cool. The signal produced by an animal calling in the forest will be reflected back down into the forest by the warm air above, increasing the intensity of the call at certain distances from the source (Wiley and Richards, 1978).

13.3   Modeling techniques

13.3.1 Spatially explicit models

Simulation models that explicitly treat the effects of environmental heterogeneity on the propagation of sound are uncommon. While much research has focused on the physiology of animal sound, behavioral aspects of signalling (e.g. Bradbury and Vehrencamp, 1998; Hopp *et al.*, 1998), and techniques for measuring the propagation of sound (e.g. Dantzker *et al.*, 1999), there are substantial opportunities for development of new models of acoustical relationships. Such models might, for example, be used to explore the relationships between animal movement patterns, the distribution of acoustical obstructions (i.e. trees, hills), and territorial calls. One such model, called "Peeper" (Parris, 2002), is used as the example model for this chapter and is described in detail below. Similarly, the spatially explicit effect of road networks and engine noise in a matrix of differing vegetation and topography might provide useful information for environmental impact analysis. While roads are frequently treated as physical barriers to movement, they are less often treated explicitly as sources of noise (Fig. 13.6).

The "Interactive Sound Information System" (ISIS) has been developed to provide information to communities and decision makers to help them to

FIGURE 13.6
Sound attenuation is affected by heterogeneous space. Sound from a snowmobile
engine is dampened by vegetation as it propagates away from a snowy trail.

understand and resolve noise management problems in spatially explicit urban
settings (Dubbink, 2000). It presents real examples of noises, such as those produced by traffic on freeways or aircraft flying overhead, by playing digital sound
recordings while displaying three-dimensional visualizations of streetscapes,
landscapes, and the traffic (or aircraft) in question. It also provides parameters
such as accumulated noise over a 24 hour period and enables the user to investigate the effect on noise levels of measures to mitigate sound (Dubbink, 2000).
Users can explore noise levels at specific locations in a human landscape by,
for example, pointing at a particular building and hearing the noise level at
that place. At this time, ISIS does not include examples of biological sounds
or the effect of human sounds on animals, but one can imagine extending the
algorithms for use in natural or semi-natural settings.

Researchers in the Bioacoustics Research Program at Cornell University have
used spatial arrays of microphones, called an automatic location system (ALS),
to study the interaction of multiple bird calls (McGregor et al., 1997; Burt, 2000).
Traditionally, playback techniques have been used to study the behavior of individual birds responding to recorded calls of other birds. Such work, although
valuable, fails to consider realistic natural conditions, where numerous birds
in different neighborhood locations call to one another and change their calls
depending on what they are hearing (McGregor et al., 1997; Burt, 2000). By
measuring the time at which a call is recorded by microphones at different locations, researchers can calculate the location of the caller and plot it on a map.

Experiments such as these are first steps for exploring the importance of spatial interaction among signalling birds.

13.4 Introduction to the example model

The northern spring peeper *Pseudacris crucifer crucifer* is a small brown to gray frog (3.7 cm in length), with a range extending through eastern North America from Nova Scotia to the Carolinas (Harding, 1997). Spring peepers breed in temporary and permanent ponds, marshes, and ditches, and choruses of calling males can be heard at breeding sites in early spring. The advertisement call of the northern spring peeper is a high-pitched peep given about once every second (Harding, 1997). This call is designed to attract conspecific females for mating and may also serve to regulate spacing between males in a chorus (Brenowitz *et al.*, 1984).

Wilczynski *et al.* (1984) analyzed the call of the northern spring peeper and investigated the physiological characteristics of the peripheral auditory system of males and females of the species. They determined that the call of the peeper was a simple, nearly tonal signal with a mean dominant frequency of 2895 Hz (range 2588–3212 Hz). The threshold sound pressure level (SPL) at which the frequency of the advertisement call could be detected differed considerably between males and females. Among males, the lowest threshold to the mean dominant frequency of the advertisement call was 78.5 dB SPL, while for females the lowest threshold to this frequency was 58 dB SPL (Wilczynski *et al.*, 1984). Thus, the advertisement call of the northern spring peeper is audible to females at much greater distances from the caller than it is to males. The amplitude of the call of the northern spring peepers is uniform around the caller (i.e., the call is not directional; Gerhardt (1975)). Following experiments in a grassland habitat with sparsely scattered trees and shrubs near Ithaca, New York, Brenowitz *et al.* (1984) derived two equations to describe the attenuation of the advertisement call of the northern spring peeper. The first equation describes attenuation of the call when the calling frog is on the ground, while the second describes attenuation when it is perched 50 cm above the ground. Spring peepers call from ground level or from elevated perches on grass, shrubs, or trees (Brenowitz *et al.*, 1984).

A spatially explicit model of propagation of an advertisement call of the spring peeper was created (Parris, 2002) using information on call characteristics and hearing thresholds from Wilczynski *et al.* (1984), the equations of Brenowitz *et al.* (1984) describing call attenuation, and information on attenuation of sound over soft and hard ground surfaces from Forrest (1994). The model has a graphical user interface in ArcView that presents a hypothetical landscape representing a grassland containing ponds of varying size. The user

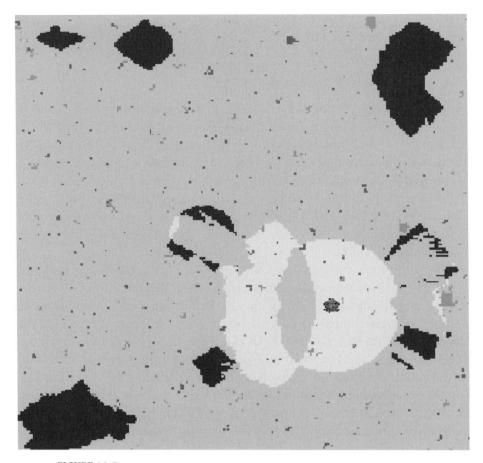

FIGURE 13.7
An example of output from the Peeper model (Parris, 2002). Sound is attenuated less
over water than when it passes over vegetated surfaces, altering the extent of
detectability.

selects a location in the landscape for the calling frog, its height above the
ground (0 or 50 cm), the amplitude of the call, and the hearing thresholds of
the male and female conspecifics. The location chosen will determine the proportion of soft surface (grass-covered soil) and hard surface (water) over which
the call will travel.

If the frog is calling from the ground, the first equation of Brenowitz *et al.*
(1984) is used to calculate attenuation as the call propagates over the soil. An
equation describing attenuation over water (the water equation), derived from
the equations of Brenowitz *et al.* (1984) and data presented by Forrest (1994), is
used when the call propagates over a pond. If the frog is calling from an elevation of 50 cm, the second equation of Brenowitz *et al.* (1984) is used to calculate

attenuation over soil, and the water equation is again used to calculate attenuation over ponds. Data presented by Forrest (1994) indicated that frequencies in the 1–4 kHz range, propagating over a hard boundary such as water, attenuate at a rate of 6 dB lower than that expected from spherical spreading. In addition, elevating the source of a sound has no effect on the rate of attenuation over water. A C program external to ArcView calculates sound propagation from the source and the shape and location of the area over which the call can be detected by females and other males in the chorus. This effective call area is then displayed on the landscape in ArcView. Users can explore the effects of caller location and elevation, call intensity, male and female hearing thresholds, and land/water surface differences on the effective call area of the northern spring peeper (Fig. 13.7).

13.5 Summary

Animal calls that attract mates, delimit territory, or warn of danger are examples of information transport by the sound vector. More passive, but also ecologically important, are sounds made by predators and prey as they move and environmental sounds, including the rumble of approaching storms or the rush of flowing water. All of these provide information to the animals that detect them. Sound propagates through the environment as waves, radiating outward from their source, with intensity decreasing as the square of travel distance. However, sound propagation is complicated by environmental heterogeneity, including surface differences, obstructions, and temperature gradients. Simulation models that explicitly consider spatial heterogeneity in natural environments are rare but represent opportunities for exploring interactions within and among animal species, between animals and their environments, and, importantly, between animals and humans.

References

Abbot, M. B. and J. C. Refsgaard (ed.) 1996. *Distributed Hydrological Modeling*. Dordrecht: Kluwer Academic.

Abbot, M. B., J. C. Bathurst, J. A. Cunge, P. E. O'Connell and J. Rasmussen. 1986. An introduction to the European Hydrological System–Système Hydrologique Européen, 'SHE' 1, History and philosophy of a physically based, distributed modeling system. *Journal of Hydrology*, **87**, 45–59.

Aber, J. D., I. C. Burke, B. Acock *et al*. 1999. Group report: hydrological and biogeochemical processes in complex landscapes – what is the role of temporal and spatial ecosystem dynamics. In *Integrating Hydrology, Ecosystem Dynamics, and Biogeochemistry in Complex Landscapes*, ed. J. D. Tenhunen and P. Kabat, pp. 335–355. Chichester, UK: Wiley.

Abrahams, A. D., A. J. Parsons and J. Wainwright. 1994. Resistance to overland flow on semiarid grassland and shrubland hillslopes, Walnut Gulch, southern Arizona. *Journal of Hydrology*, **156**, 431–446.

Abrahams, A. D., G. Li and A. J. Parsons. 1996. Rill hydraulics on a semiarid hillslope, southern Arizona. *Earth Surface Processes and Landforms*, **21**, 35–47.

Abtew, W., J. M. Gregory and J. Borrelli. 1989. Wind profile: estimation of displacement height and aerodynamic roughness. *Transactions of the American Society of Agricultural Engineers*, **32**, 521–527.

Albini, F. A. 1979. *Spot fire distance from burning trees: a predictive model. General Technical Report INT-56*. Ogden, UT: USDA Forest Service.

Allen, T. F. H. 1998. The landscape "level" is dead: persuading the family to take it off the respirator. In *Ecological Scale. Theory and Application*, ed. D. L. Peterson and V. T. Parker, pp. 35–54. New York: Columbia University Press.

Allen, T. F. H. and T. W. Hoekstra. 1990. The confusion between scale-defined levels and conventional levels of organization in ecology. *Journal of Vegetation Science*, **1**, 5–12. 1992. *Toward a Unified Ecology*. New York: Columbia University Press.

Allen, T. F. H., R. V. O'Neill and T. W. Hoekstra. 1987. Interlevel relations in ecological research and management: some working principles from hierarchy theory. *Journal of Applied Systems Analysis*, **14**, 63–79.

Anderson, D. G., E. A. Catchpole, N. J. DeMestre and T. Parkes. 1982. Modeling the spread of grass fires. *Journal of the Australian Mathematical Society (Series B)*, **23**, 451–466.

Anderson, H. E. 1983. *Predicting Wind-driven Wild Land Fire Size and Shape. Research Paper INT-305*. Ogden, UT: USDA Forest Service, Intermountain Forest and Range Experiment Station.

Anderson, H. W. 1954. Suspended sediment discharge as related to streamflow, topography, soil, and land use. *Transactions of the American Geophysical Union*, **35**, 268–281.

1981. Sources of sediment-induced reductions in water quality appraised from catchment attributes and land use. *Journal of Hydrology*, **51**, 347–358.

Anderson, L., C. E. Carlson and R. H. Wakimoto. 1987. Forest fire frequency and western spruce budworm in western Montana. *Forest Ecology and Management*, **22**, 251–260.

Anderson, M. P. 1995. Groundwater modeling in the 21st century. In *Groundwater Models for Resources Analysis and Management*, ed. A. I. El-Kadi, pp. 79–93. Boca Raton, FL: Lewis.

Anderson, M. P., and W. W. Woessner. 1992. *Applied Groundwater Modeling. Simulation of Flow and Advective Transport*. San Diego, CA: Academic Press.

Andreae, M. O. 1991. Biomass burning: its history, use and distribution and its impact on environmental quality and global climate. In *Global Biomass Burning: Atmospheric, Climatic and Biospheric Implications*, ed. J. S. Levine, pp. 3–21. Cambridge, MA: MIT Press.

Andrews, P. L. 1986. *BEHAVE: Fire Behavior Prediction and Fuel Modeling System BURN Subsystem, Part 1. General Technical Report INT-194*. Ogden, UT: USDA Forest Service.

Andrews, P. L. and C. H. Chase. 1989. *Fire behavior prediction and fuel Modeling System BURN Subsystem, Part 2. General Technical Report INT-260*. Ogden, UT: USDA Forest Service.

Appelo, C. A. J. and D. Postma. 1996. *Geochemistry, Groundwater and Pollution*. Rotterdam: A. A. Balkema.

Archer, D. E., G. Eshel, A. Winguth *et al.*, 2000. Atmospheric P_{CO_2} sensitivity to the biological pump in the oceans. *Global Biogeochemical Cycles*, **14**, 1219–1230.

Archibold, O. W. 1995. *Ecology of World Vegetation*. London: Chapman Hall.

Arnold, J. G., P. M. Allen and G. Bernhardt. 1993. A comprehensive surface–groundwater flow model. *Journal of Hydrology*, **142**, 47–69.

Avissar, R. 1992. Conceptual aspects of a statistical–dynamical approach to represent landscape subgrid-scale heterogeneities in atmospheric models. *Journal of Geophysical Research*, **97**(D3), 2729–2742.

Axtmann, E. V. and S. N. Luoma. 1991. Large-scale distribution of metal contamination in the fine-grained sediments of the Clark Fork River, Montana, USA. *Applied Geochemistry*, **6**, 75–88.

Aylor, D. E. 1990. The role of intermittent wind in the dispersal of fungal pathogens. *Annual Review of Phytopathology*, **28**, 73–92.

Aylor, D. E. and J.-Y. Parlange. 1975. Ventilation required to entrain small particles from leaves. *Plant Physiology*, **56**, 97–99.

Aylor, D. E., H. A. McCarty and A. Bainbridge. 1981. Deposition of particles liberated in gusts of wind. *Journal of Applied Meteorology*, **20**, 1212–1221.

Bachmat, Y. and J. Bear. 1964. The general equations of hydrodynamic dispersion in isotropic porous medium. *Journal of Geophysical Research*, **69**, 2561–2567.

Bagnold, R. A. 1941. *The Physics of Blown Sand and Desert Dunes*. London: Methuen.

1954. Experiments on a gravity-free dispersion of large solid spheres in a Newtonian fluid under shear. *Proceedings of the Royal Society of London, Series A*, **225**, 49–63.

1988. *The Physics of Sediment Transport by Wind and Water: A Collection of Hallmark Papers*. Reston, VA, American Society of Civil Engineers.

Baker, R. R. 1978. *The Evolutionary Ecology of Animal Migration*. New York: Holmes and Meier.

Baker, W. L. 1992. Effect of settlement and fire suppression on landscape structure. *Ecology*, **73**, 1879–1887.

Barber, R. T. 1988. Ocean basin ecosystems. In *Concepts of Ecosystem Ecology*, ed. L. L. Pomeroy and J. J. Alberts, pp. 171–193. New York: Springer-Verlag.

Barbour, M. G., J. H. Burk and W. D. Pitts. 1987. *Terrestrial Plant Ecology*, 2nd edn. Menlo Park, CA: Benjamin/Cummings.

Barca, D., S. Di Gregario, F. P. Nicoletta and M. Sorriso Valvo. 1986. A cellular space model for flow type landslides. In *Proceedings of the IASTED International Symposium on Computers and their Application for*

Development, ed. G. Messina and M. H. Hamzda, Taormina, Italy, September 1986.

Barnard, J. 2002. Closer look at wildfire shows that not all is black. *The Register-Guard* (Eugene, OR), September 21.

Baumgartner, M. F. and G. Apfl. 1994. Towards an integrated geographic analysis system with remote sensing, GIS, and consecutive modeling for snow cover monitoring. *International Journal of Remote Sensing,* **15,** 1507–1517.

Beasley, D. B. and L. F. Huggins. 1982. *ANSWERS (Areal Nonpoint Source Watershed Environmental Response Simulation) User's Manual. Report 905/9-82-001.* Chicago, IL: US Environmental Protection Agency.

Begon, M., J. L. Harper and C. R. Townsend. 1986. *Ecology. Individuals, Populations, and Communities.* Sunderland, MA: Sinauer.

Benda, L. and T. W. Cundy. 1990. Predicting deposition of debris flows in mountain channels. *Canadian Geotechnical Journal,* **27,** 409–417.

Benda, L. and T. Dunne. 1997. Stochastic forcing of sediment supply to channel networks from landsliding and debris flow. *Water Resources Research,* **33,** 2849–2863.

Benda, L., T. J. Beechie, A. Johnson and R. C. Wissmar. 1992. The geomorphic structure of salmonid habitats in a recently deglaciated river basin, Washington state. *Canadian Journal of Fisheries and Aquatic Sciences,* **49,** 1246–1256.

Bendell, J. F. 1974. Effects of fire on birds and mammals. In *Fire and Ecosystems,* ed. T. T. Kozlowski and C. E. Ahlgren, pp. 73–138. New York: Academic Press.

Benedict, J. B. 1993. Influence of snow upon granidiorite weathering, Colorado Front Range, USA. *Boreas,* **22,** 87–92.

Benfield, M. C., P. H. Wiebe, T. K. Stanton *et al.* 1998. Estimating the spatial distribution of zooplankton biomass by combining video plankton recorder and single-frequency acoustic data. *Deep-Sea Research II,* **45,** 1175–1199.

Berendsen, M. E. and W. A. Reiners. 1999. *A GIS-based decision support tool for assessing BLM permit applications.* Report to the Wyoming State Office of the Bureau of Land Management, Cheyenne, WY.

Berner, E. K. and R. A. Berner. 1996. *Global Environment. Water, Air and Geochemical Cycles.* New Jersey: Prentice Hall.

Beven, K. 1996. A discussion of distributed hydrological modelling. In *Distributed Hydrological Modelling,* ed. M. B. Abbott and J. C. Refsgaard, pp. 255–278. Dordrecht: Kluwer Academic.

Beven, K. and P. Germann. 1982. Macropores and water flow in soils. *Water Resources Research,* **18,** 1311–1325.

Beven, K. and I. D. Moore. 1993. *Terrain Analysis and Distributed Modelling in Hydrology.* Chichester, UK: Wiley.

Bicknell, B. R., J. C. Imhoff, J. L. Kittle, Jr, A. S. Donigian and R. C. Johanson. (1993). *Hydrologic Simulation Program – FORTRAN (HSPF): Users Manual for Release 10. Report 600/R-93/174.* Athens, GA: US Environmental Protection Agency.

Billings, W. D. 1969. Vegetational pattern near alpine timberline as affected by fire-snowdrift interactions. *Vegetatio,* **19,** 192–207.

Billings, W. D. and L. C. Bliss. 1959. An alpine snowbank environment and its effects on vegetation, plant development, and productivity. *Ecology,* **40,** 388–397.

Boone, R. B. and M. L. Hunter, Jr. 1996. Using diffusion models to simulate the effects of land use on grizzly bear dispersal in the Rocky Mountains. *Landscape Ecology,* **11,** 51–64.

Boose, E. R., K. E. Chamberlin and D. R. Foster. 2001. Landscape and regional impacts of hurricanes in New England. *Ecological Monographs,* **71,** 27–48.

Bormann, F. H. and G. E. Likens. 1979. Catastrophic disturbance and the steady state in northern hardwood forests. *American Scientist:* **67,** 660–669.

Bovet, P. and S. Benhamou. 1988. Spatial analysis of animals' movements using a correlated random walk model. *Journal of Theoretical Biology,* **131,** 419–433.

Bradbury, J. W. and S. L. Vehrencamp. 1998. *Principles of Animal Communication.* Sunderland, MA: Sinauer.

Bradshaw, G. A. and T. A. Spies. 1992. Characterizing canopy gap structure in forests using wavelet analysis. *Journal of Ecology*, **80**, 205–215.

Brenowitz, E. A. 1986. Environmental influences on acoustic and electric animal communication. *Brain, Behavior and Evolution*, **28**, 32–42.

Brenowitz, E. A., W. Wilczynski and H. H. Zakon. 1984. Acoustic communication in spring peepers: environmental and behavioral aspects. *Journal of Comparative Physiology*, **155**, 585–592.

Bristow, K. L. and G. S. Campbell. 1984. On the relationship between incoming solar radiation and daily maximum and minimum temperature. *Agricultural and Forest Meteorology*, **31**, 159–166.

Broecker, W. S. 1986. *How to Build a Habitable Planet*. Palisades, NY: Eldigio Press.

  1987. Unpleasant surprises in the greenhouse? *Nature*, **328**, 123–126.

  1997. Thermohaline circulation, the Achilles heel of our climate system: will man-made CO_2 upset the current balance? *Science*, **278**, 1582–1588.

Brooks, J. L. and S. I. Dodson. 1965. Predation, body size, and composition of plankton. *Science*, **150**, 28–35.

Brooks, P. D., M. W. Williams, D. A. Walker and S. K. Schmidt. 1995. The Niwot Ridge snow fence experiment: biogeochemical responses to changes in the seasonal snow pack. In *Proceedings of a Symposium on Biogeochemistry of Seasonally Snow-Covered Catchments*, Boulder, CO, July 1995, IAHS Publication No. 228, pp. 293–302. Wallingford, UK: International Association of Hydrological Sciences.

Brooks, P. D., S. K. Schmidt and M. W. Williams. 1997. Winter production of CO_2 and N_2O from alpine tundra: environmental controls and relationship to inter-system C and N fluxes. *Oecologia*, **110**, 403–413.

Brown, G. C. and C. E. Cooper. 1995. *Bioenergetics: A Practical Approach*. Oxford University Press, New York.

Brown, J. H. 1995. *Macroecology*. Chicago, IL: University of Chicago Press.

Brown, J. H. and A. C. Gibson. 1983. *Biogeography*. St Louis, MO: Mosby.

Brozovic, N., D. W. Burbank, and A. J. Meigs. 1997. Climatic limits on landscape development in the Northwestern Himalaya. *Science*, **276**, 571–574.

Brunsden, D. and D. B. Prior. 1984. *Slope Instability*. Chichester, UK: Wiley.

Buchan, L. A. J. and D. K. Padilla. 1999. Estimating the probability of long-distance overland dispersal of invading aquatic species. *Ecological Applications*, **9**, 254–265.

Buchman, S. L. and G. P. Nabhan. 1996. *The Forgotten Pollinators*. Washington, DC: Island Press.

Bugbee, B. G. 1985. Calculating CO_2 fluxes into leaves. In *Plant Physiology*, 3rd edn, ed. F. B. Salisbury and C. W. Ross, pp. 64–65. Belmont, CA: Wadsworth.

Burke, I. C. 2000. Landscape and regional biogeochemistry: approaches. In *Methods in Ecosystem Science*, ed. O. E. Sala and H. A. Mooney, pp. 277–288. New York: Springer-Verlag.

Burleigh, J. R., R. L. Edmonds, J. B. Harrington, Jr *et al.* 1979. Modeling of aerobiological systems. In *Aerobiology. The Ecological Systems Approach (US/IBP Synthesis Series No. 10)*, ed. R. L. Edmonds, pp. 279–371. Stoudsburg, PA: Dowden, Hutchinson and Ross.

Burrough, P. A. and R. A. McDonnell. 1998. *Principles of Geographical Information Systems*. Oxford: Oxford University Press.

Burt, J. M. 2000. Use of a radio microphone array to study banded wren song interactions at the neighborhood level. *Journal of the Acoustical Society of America*, **108**, 2583.

Burt, W. H. 1943. Territoriality and home range concepts as applied to mammals. *Journal of Mammology*, **24**, 346–352.

Buskirk, S. W. and R. A. Powell. 1994. Habitat ecology of fishers and American martens. In *Martens, Sables and Fishers: Biology and Conservation*, ed. S. W. Buskirk, A. S. Harestad, M. G. Raphael and R. A. Powell, pp. 283–296. Ithaca, NY: Cornell University Press.

Butcher S. S. 1992. Equilibrium, rate, and natural systems. In *Global Biogeochemical Cycles*, ed. S. S. Butcher, R. J. Charlson, G. H. Orians, and G. V. Wolfe, pp. 73–92. San Diego, CA: Academic Press.

Campbell, G. S. 1977. *An Introduction to Environmental Biophysics*. New York: Springer-Verlag.

1985. *Soil Physics with Basic Transport Models for Soil–Plant Systems*. Amsterdam: Elsevier Science.

Campbell, G. S. and J. M. Norman. 1998. *Introduction to Environmental Biophysics*, 2nd edn. New York: Springer-Verlag.

Campbell, N. A. and J. B. Reece. 2001. *Biology*, 6th edn. Redwood City, CA: Benjamin/Cummings.

Campbell, N. A., J. B. Reece and L. G. Mitchell. 1999. *Biology*, 5th edn. Menlo Park, CA: Benjamin/Cummings.

Canale, R. P., L. M. DePalma, and A. H. Vogel. 1976. A plankton-based food web model for Lake Michigan. In *Modeling Biochemical Processes in Aquatic Ecosystems*, ed. R. P. Canale, pp. 33–77. Ann Arbor, MI: Ann Arbor Science.

Carlquist, S. 1981. Chance dispersal. *American Scientist*, **69**, 509–515.

Carroll, J. J. and D. R. Viglierchio. 1981. On the transport of nematodes by the wind. *Journal of Nematology*, **13**, 476–483.

Carson, M. A. 1969. Models of hillslope development under mass failure. *Geographical Analysis*, **1**, 76–100.

Caswell, H. 1978. Predator-mediated coexistence: a non-equilibrium model. *American Naturalist*, **112**, 127–154.

Chadwick, A. J. and J. C. Morfett. 1998. *Hydraulics in Civil and Environmental Engineering*, 3rd edn. London: Chapman & Hall.

Chadwick, O. A., L. A. Derry, P. M. Vitousek, B. J. Huebert and L. O. Hedin. 1999. Changing sources of nutrients during four million years of ecosystem development. *Nature*, **397**, 491–497.

Chandler, D. 1998. Finding transition pathways: throwing ropes over rough mountain passes, in the dark. In *Computer Simulation of Rare Events and Dynamics of Classical and Quantum Condensed Phase Systems: Classical and Quantum Dynamics in Condensed Phase Simulations*, ed. B. J. Berne, G. Ciccotti and D. F. Coker, pp. 51–66. Singapore: World Scientific.

Chapin, F. S., III. 1980. The mineral nutrition of wild plants. *Annual Review of Ecology and Systematics*, **11**, 233–260.

Chapin, F. S., III, P. A. Matson and H. A. Mooney. 2002. *Principles of Terrestrial Ecosystem Ecology*. New York: Springer-Verlag.

Chapin, T. G., D. J. Harrison and D. D. Katnik. 1997. Influence of landscape pattern on habitat use by American Marten in an industrial forest. *Conservation Biology*, **12**, 1327–1337.

Chapra, S. C. 1997. *Surface Water-quality Modeling*. (McGraw-Hill Series in Water Resources and Environmental Engineering.) New York: McGraw-Hill.

Chapra, S. C. and K. H. Reckhow. 1983. *Engineering Approaches for Lake Management*, Vol. 2: *Mechanistic Modeling*. Woburn, MA: Butterworth.

Charles-Dominique, P. 1986. Inter-relations between frugivorous vertebrates and pioneer plants: Cecropia, birds and bats in French Guyana. In *Frugivores and Seed Dispersal. Tasks for Vegetation Science 15*, ed. A. Estrada, and T. H. Fleming, pp. 119–135. Dordrecht: W. Junk.

Chatigny, M. A., R. L. Dimmick and J. B. Harrington. 1979. Deposition. In *Aerobiology. The Ecological Systems Approach* (US/IBP Synthesis Series No. 10), ed. R. L. Edmonds, pp. 111–150. Stoudsburg, PA: Dowden, Hutchinson and Ross.

Chen, C. W. 1970. Concepts and utilities of ecological models. *Journal of Sanitary Engineering Division, ASCE*, **96**, 1085–1086.

Choy, B. and D. D. Reible. 2000. *Diffusion Models of Environmental Transport*. Boca Raton, FL: Lewis.

Chung, C. F. and A. G. Fabbri. 1999. Probabilistic prediction models for landslide hazard mapping. *Photogrammetric Engineering and Remote Sensing*, **65**, 1389–1399.

Chuvieco, E. and R. G. Congalton. 1989. Application of remote sensing and geographic information systems to forest fire hazard mapping. *Remote Sensing of Environment*, **29**, 147–159.

Clark, J. S., C. Fastie, G. Hurtt *et al*. 1998. Reid's paradox of rapid plant migration. *BioScience*, **48**, 13–24.

Clarke, K. C. 1997. *Getting Started with Geographic Information Systems*. Upper Saddle River, NJ: Prentice Hall.

  2001. *Getting Started With Geographic Information Systems*, 3rd edn. Upper Saddle River, NJ: Prentice Hall.

Clarke, K. C. and G. Olsen. 1996. Refining a cellular automaton model of wildfire propagation and extinction. In *GIS and Environmental Modeling: Progress and Research Issues*, ed. M. F. Goodchild, L. T. Steyaert, B. O. Parks *et al.*, pp. 333–338. Fort Collins, CO: GIS World Books.

Clarke, K. C., J. A. Brass and P. A. Riggan. 1994. Refining a cellular automaton model of wildfire propagation and extinction. *Photogrammetric Engineering and Remote Sensing*, **60**, 1355–1367.

Clarke, K. C., B. O. Parks and M. P. Crane. 2002. *Geographic Information Systems and Environmental Modeling*. Upper Saddle River, NJ: Prentice Hall.

Clements, F. E. 1916. *Plant Succession: An Analysis of the Development of Vegetation, Publication 242*. Washington, DC: Carnegie Institute.

Clements, F. E. and V. E. Shelford. 1939. *Bio-ecology*. New York: Wiley.

Collins, S. L. 1992. Fire frequency and community heterogeneity in tallgrass prairie vegetation. *Ecology*, **73**, 2001–2006.

Cooper, W. S. 1913. The climax forest of Isle Royale, Lake Superior, and its development. *Botanical Gazette*, **55**, 1–44, 115–140, 189–235.

Cornet, A. F., C. Montana, J. P. Delhoume and J. Lopez-Portillo. 1992. Water flows and the dynamics of desert vegetation stripes. In *Landscape Boundaries: Ecological Consequences on Biotic Diversity and Ecological Flows*, ed. A. J. Hansen and F. di Castri, pp. 327–345. New York: Springer-Verlag.

Costa-Cabral, M. C. and S. J. Burges. 1994. Digital elevation model networks (DEMON): a model of flow over hillslopes for computation of contributing and dispersal areas. *Water Resources Research*, **30**, 1681–1692.

Cottam, G. and R. P. McIntosh. 1966. Vegetational continuum. *Science*, **152**, 546–547.

Couclelis, H. 1999. Space, time, geography. In *Geographical Information Systems*, 2nd edn, ed. P. Longley, M. F. Goodchild, D. J. Maquire and D. W. Rhind, pp. 29–38. New York: Wiley.

Coulson, S. J., I. D. Hodkinson, A. T. Strathdee *et al.* 1995. Thermal environments of arctic soil organisms during winter. *Arctic and Alpine Research*, **27**, 364–370.

Cowen, R. K., K. M. M. Lwiza, S. Sonaugle, C. B. Paris and D. B. Olson. 2000. Connectivity of marine populations: open or closed? *Science*, **287**, 857–859.

Cowles, H. C. 1899. The ecological relations of the vegetation on the sand dunes of Lake Michigan. *Botanical Gazette*, **27**, 95–117, 167–175.

  1901. The physiographic ecology of Chicago and vicinity. *Botanical Gazette*, **31**, 73–108, 145–181.

Crist, T. O., D. S. Guertin, J. A. Wiens and B. T. Milne. 1992. Animal movement in heterogeneous landscapes: an experiment with *Eleodes* beetles in shortgrass prairie. *Functional Ecology*, **6**, 536–544.

Crutzen, P. J. and M. O. Andreae. 1990. Biomass burning in the tropics: impact on atmospheric chemistry and biogeochemical cycles. *Science*, **250**, 1669–1678.

Culling, W. E. H. 1960. Analytical theory of erosion. *Journal of Geology*, **68**, 336–344.

Curtis, J. T. 1959. *The Vegetation of Wisconsin: An Ordination of Plant Communities*. Madison, WI: University of Wisconsin Press.

Cuthill, I. C. and A. I. Houston. 1997. Managing time and energy. In *Behavioral Ecology. An Evolutionary Approach*, 4th edn, ed. J. R. Krebs and N. B. Davies, pp. 97–120. Oxford: Blackwell Science.

Dale, V. H., M. A. Hemstrom and J. F. Franklin. 1984. The effect of disturbance frequency on forest succession in the Pacific Northwest. In *New Forests for a Changing World: Proceedings of the 1983 Convention of the Society of American Foresters*, Bethesda, October 1983, pp. 300–304. Portland, OR: Society of American Foresters.

  1986. Modeling the long-term effects of disturbances on forest succession, Olympic Peninsula, Washington. *Canadian Journal of Forest Research*, **16**, 56–67.

D'Antonio, C. M. and P. M. Vitousek. 1992. Biological invasions by exotic grasses, the grass/fire cycle, and global change. *Annual Review of Ecology and Systematics*, **23**, 63–87.

Dantzker, M. S., G. B. Deane and J. W. Bradbury. 1999. Directional acoustic radiation in the strut display of male sage grouse *Cantrocercus urophasianus*. *Journal of Experimental Biology*, **202**, 2893–2909.

Darwin, C. 1889. *The Various Contrivances by which Orchids are Fertilized by Insects*. New York: Appleton.

Daubechies, I. 1992. Ten lectures on wavelets. *CBMS-NSF Regional Conference Series in Applied Mathematics*. Philadelphia, PA: Society for Industrial and Applied Mathematics. [Seen in Keitt, 2000.]

Daubenmire, R. 1966. Vegetation: identification of typal communities. *Science*, **151**, 291–298.

David, C. T., J. S. Kennedy, A. R. Ludlow, J. N. Perry and C. Wall. 1982. A reappraisal of insect flight toward a distant, point-source of wind-borne odor. *Journal of Chemical Ecology*, **9**, 1207–1215.

Davidson, E. A. and L. V. Verchot. 2000. Testing the hole-in-pipe model of nitric and nitrous oxide emissions from soils using the TRAGNET database. *Global Biogeochemical Cycles*, **14**, 1035–1043.

Davidson, E. A., M. Keller, H. E. Erickson, L. V. Verchot and E. Veldkamp. 2000. Testing a conceptual model of soil emissions of nitrous and nitric oxides. *BioScience*, **50**, 667–680.

Davis, F. W. and D. A. Burrows. 1994. Spatial simulation of fire regimes in Mediterranean-climate landscapes. In *The Role of Fire in Mediterranean-type Ecosystems*, ed. J. M. Moreno and W. C. Oechel, pp. 117–139. Berlin: Springer-Verlag.

Davis, M. B. 1986. Climatic stability, time lags, and community disequilibrium. In *Community Ecology*, ed. J. Diamond and T. J. Case, pp. 269–284. New York: Harper and Row.

Day, K. 2000. Aquifer Heterogeneity in Groundwater Flow Modeling. M.Sc. Thesis. Department of Geology and Geophysics, University of Wyoming, Laramie, WY.

De Floriani, L. and P. Magillo 1999. Intervisibility on terrains. In *Geographical Information Systems: Principles, Techniques, Management and Applications*, 2nd edn, ed. P. A. Longley, M. F. Goodchild, D. J. Maguire and D. W. Rhind. Chichester, UK: Wiley.

De Jong, M. D., P. S. Wagenmakers and J. Goudriaan. 1991. Modelling the escape of *Chondrostereum purpureum* spores from a larch forest with biological control of *Prunus serotina*. *Netherlands Journal of Plant Pathology*, **97**, 55–61.

De Ploey, J. 1990. Threshold conditions for thalweg gullying with special reference to loess areas. In *Soil Erosion: Experiments and Models (Catena Supplement 17)*, ed. R. B. Bryan, pp. 147–151. Reiskirchen, Germany: Catena Verlag.

De Roo, A. P. J., Hazelhoff, L. and Burrough, P. A. 1989. Soil erosion modelling using ANSWERS and geographical information systems. *Earth Surface Processes and Landforms*, **14**, 517–532.

de Vries, M. 1993. *Use of Models for River Problems*. *Studies and Reports in Hydrology* 51. Paris: UNESCO.

De Wispelaere, C. (ed.) 1981. *Proceedings of the International Technical Meeting on Air Pollution Modeling and its Application*. Amsterdam: Plenum Press.

Deangeli, C., G. P. Giani and E. Segre. 1994. Modellazione Numerica di Colate Detritiche con il Metodo degli Automi Cellulari. In *Proceedings of the Conference: Ruolo dei Fluidi nei Problemi di Ingegneria Geotecnica*, Mondovi, Italy, Vol. 1.

DeAngelis, D. L. and G. T. Yeh. 1984. An introduction to modeling migratory behavior of fishes. In *Mechanisms of Migration in Fishes*, ed. J. D. McCleave, G. P. Arnold, J. J. Dodson and W. H. Neill, pp. 445–469. New York: Plenum Press.

Deeming, J. E., R. E. Burgan and J. D. Cohen. 1977. *General Technical Report INT-39: The National Fire-danger Rating System 1978*. Ogden, UT: USDA Forest Service Intermountain Research Station.

Del Grosso, S. J., W. J. Parton, A. R. Mosier *et al.* 2000. General model for N_2O and N_2 gas emissions from soils due to denitrification. *Global Biogeochemical Cycles*, **14**, 1045–1060.

Dickinson, R. E., A. Henderson-Sellers, P. J. Kennedy and M. R. Wilson. 1986. *Biosphere Atmosphere Transfer Scheme for the NCAR Community Climate Model. NCAR Technical Note NCAR/TN-275+STR.* Boulder, CO: National Center for Atmospheric Research.

Dickinson, R. E., A. Henderson-Sellers and P. J. Kennedy. 1993. *Biosphere–Atmosphere Transfer Scheme (BATS) Version 1e as Coupled to the NCAR Community Climate Model. NCAR Technical Note TN-387+STR.* Boulder, CO: National Center for Atmospheric Research.

Di Gregorio, S., R. Rongo, C. Siciliano, M. Sorriso-Valvo and W. Spataro. 1999. Mount Ontake landslide simulation by the cellular automata model SCIDDICA-3. *Physics and Chemistry of the Earth*, **24**, 131–137.

Dingle, H. 1980. Ecology and evolution of migration. In *Animal Migration, Orientation and Navigation*, ed. S. A. Gauthreaux, Jr, pp. 1–2, 78–83. New York: Academic Press.

Dobzhansky, T., J. R. Powell, C. E. Taylor and M. Andregg. 1979. Ecological variables affecting the dispersal behavior of *Drosophila pseudoobscura* and its relatives. *American Naturalist*, **114**, 325–334.

Donohue, D. 1986. Review and analysis of slow mass movement mechanisms with reference to a Weardale catchment, N. England. *Zeitschrift für Geomorphologie Supplementband*, **60**, 41–54.

Driese, K. L. and W. A. Reiners. 1998. Aerodynamic roughness parameters for semi-arid natural shrub communities of Wyoming, USA. *Agricultural and Forest Meteorology*, **88**, 1–14.

Driese, K. L., W. A. Reiners, E. H. Merrill and K. G. Gerow. 1997. A digital land cover map of Wyoming, USA: a tool for vegetation analysis. *Journal of Vegetation Science*, **8**, 133–146.

Dubbink, D. T. 2000. Visualizations for noise management. In *Proceedings of the 2000 APA National Planning Conference, New York*.

Duecker, K. J. 1979. Land resource information systems: a review of fifteen years experience. *Geo-Processing*, **17**, 60–76. [Seen in Clarke, 1997.]

Duellman, W. E. and L. Trueb. 1986. *Biology of Amphibians.* New York: McGraw-Hill.

Dunning, J. B., Jr, D. J. Stewart, B. J. Danielson *et al.* 1995. Spatially explicit population models: current forms and future uses. *Ecological Applications*, **5**, 3–11.

Dusenbery, D. B. 1992. *Sensory Ecology: How Organisms Acquire and Respond to Information.* New York: Freeman.

Dyer, A. G. 1998. The colour of flowers in spectrally variable illumination and insect pollinator vision. *Journal of Comparative Physiology*, **183**, 203–212.

ECOMAP. 1993. *National Hierarchical Framework of Ecological Units.* Washington, DC: ECOMAP, USDA Forest Service.

Edmonds, R. L. 1979. Aerobiology. In *The Ecological Systems Approach. US/IGP Synthesis Series No. 10.* Stroudsburg, PA: Dowden, Hutchinson and Ross.

Egenhofer, M. J. and R. G. Golledge. 1998. *Spatial and Temporal Reasoning in Geographic Information Systems,* New York: Oxford University Press.

Eliassen, A. 1980. A review of long-range transport modeling. *Journal of Applied Meteorology*, **19**, 231–240.

El-Kadi, A. I. (ed.). 1995. *Groundwater Models for Resources Analysis and Management.* Boca Raton, FL: Lewis.

Elkinton, J. S. and R. T. Carde. 1984. Odor dispersion. In *Chemical Ecology of Insects*, ed. W. J. Bell and R. T. Carde, pp. 73–91. Sunderland, MA: Sinauer.

Elkinton, J. S., R. T. Carde and C. J. Mason. 1984. Evaluation of time-average dispersion models for estimating pheromone concentration in a deciduous forest. *Journal of Chemical Ecology*, **10**, 1081–1108.

Elkinton, J. S., Schal, C., Ono, T. and Carde, R. T. 1987. Pheromone puff trajectory and upwind flight of male gypsy moths in a forest. *Physiological Entomology*, **12**, 399–406.

Elton, C. S. 1927. *Animal Ecology.* New York: Macmillan.
   1966. *The Pattern of Animal Communities.* London: Methuen.

Ely, L. L., Y. Enzel, V. R. Baker and D. R. Cayan. 1993. A 5000-year record of extreme floods and climate change in the southwestern United States. *Science*, **262**, 410–412.

Embleton, T. F. W., J. E. Piercy and G. A. Daigle. 1983. Effective flow resistivity of ground surfaces determined by acoustical measurements. *Journal of the Acoustical Society of America*, **74**, 1239–1244.

Endler, J. A. 1993. The color of light in forests and its implications. *Ecological Monographs*, **63**, 1–27.

Endler, J. A. and M. Thery. 1996. Interacting effects of lek placement, display behavior, ambient light, and color patterns in three neotropical forest-dwelling birds. *American Naturalist*, **148**, 421–452.

EPA. 1995. *QUAL2E Windows Interface User's Manual*. Washington, DC: US Environment Protection Agency.

   2001. *Better Assessment Science Integrating Point and Nonpoint Sources. BASINS Version 3.0. User's Manual. Publication 823-B-01-001*. Washington, DC: US Environmental Protection Agency, Office of Water.

Estrada, A. and R. Coates-Estrada. 1986. Use of leaf resources by howling monkeys (*Aloutta palliata*) and leaf-cutting ants (*Atta cephalotes*) in the tropical rain forest of Los Tuxlas, Mexico. *American Journal of Primatology*, **10**, 51–66.

Estrada, A. and T. H. Fleming (ed.) 1986. *Frugivores and Seed Dispersal. Tasks for Vegetation Science 15*. Dordrecht: W. Junk.

Evans, S. G. and O. Hungr. 1993. The assessment of rockfall hazard at the base of talus slopes. *Canadian Geotechnical Journal*, **30**, 620–636.

Everest, F. H., R. L. Beschta, J. C. Scrivener *et al.* 1987. Fine sediment and salmonid production: a paradox. In *Streamside Management: Forestry and Fishery Interactions (Institute of Forest Resources Contribution No. 57)*, ed. E. O. Salo and T. W. Cundy, pp. 98–142. Seattle, WA: University of Washington.

Eyles, N. (ed.) 1983. *Glacial Geology. An introduction for Engineers and Earth Scientists*. Oxford: Pergamon Pess.

Fahrig, L. and G. Merriam. 1994. Conservation of fragmented populations. *Conservation Biology*, **8**, 50–59.

Farina, A. 1997. *Principles and Methods in Landscape Ecology*. London: Chapman & Hall.

   2000. *Landscape Ecology in Action*. Dordrecht: Kluwer Academic.

Federer, C. A. and C. B. Tanner. 1966. Spectral distribution of light in the forest. *Ecology*, **47**, 555–560.

Finney, M. A. 1993. Modeling the spread and behavior of prescribed natural fires. In *Proceedings of the 12th International Conference on Fire and Forest Meteorology*, Jekyll Island, GA, October, 1993, pp. 138–143.

   1996. *FARSITE^{TM}. Fire Area Simulator, Version 2.0. User's Guide and Technical Documentation*. Missoula, MT: Systems for Environmental Management.

Firestone, M. K. and E. A. Davidson. 1989. Microbiological basis of NO and N_2O production and consumption in soil. In *Exchange of Trace Gases between Terrestrial Ecosystems and the Atmosphere*, ed. M. O. Andreae and D. S. Schimel, pp. 7–21. New York: Wiley.

Fischer, M. M. 2000. Spatial interaction models and the role of geographic information systems. In *Spatial Models and GIS. New Potential and New Models*, ed. A. S. Fotheringham and M. Wegener, pp. 33–43. London: Taylor & Francis.

Forbes, S. A. 1925. The lake as a microcosm. *Illinois Natural History Survey Bulletin*, **15**, 537–550.

Forman, R. T. T. 1995. *Land Mosaics. The Ecology of Landscapes and Regions*. Cambridge, UK: Cambridge University Press.

Forman, R. T. T. and P. N. Moore. 1992. Theoretical foundations for understanding boundaries in landscape mosaics. In *Landscape Boundaries. Consequences for Biotic Diversity and Ecological Flow*, ed. A. J. Hansen and F. di Castri, pp. 246–258. New York: Springer-Verlag.

Forrest, T. G. 1994. From sender to receiver: propagation and environmental effects on acoustic signals. *American Zoologist*, **34**, 644–654.

Forrester, J. W. 1961. *Industrial Dynamics*. Cambridge, MA: MIT Press.

Foster, G. R. 1991. Advances in wind and water erosion prediction. *Journal of Soil and Water Conservation*, **46**, 27–29.

Foster, T. 1976. *Bushfire: History, Prevention, Control*. Sydney: Reid.

Fotheringham, A. S. 2000. GIS-based spatial modelling: a step forward or a step backwards? In *Spatial Models and GIS. New Potential and New Models*, ed. A. S. Fotheringham and M. Wegener, pp. 21–30. London: Taylor & Francis.

Fotheringham, A. S. and M. Wegener (ed.) 2000. *Spatial Models and GIS. New Potential and New Models*. London: Taylor & Francis.

Frandsen, W. H. and P. L. Andrews. 1979. *Fire behavior in non-uniform fuels. Research Paper INT-232*. Ogden, UT: USDA Forest Service Intermountain Forest and Range Experiment Station.

Frank, D. A., R. S. Inouye, N. Huntley, G. W. Minshall and J. E. Anderson. 1994. The biogeochemistry of north-temperate grassland with native ungulates: nitrogen dynamics in Yellowstone National Park. *Biogeochemistry*, **26**, 163–188.

Freeze, R. A. 1975. A stochastic conceptual analysis of one-dimensional groundwater flow in nonuniform homogeneous media. *Water Resources Research*, **11**, 725–741.

Freeze, R. A. and J. A. Cherry. 1979. *Groundwater*. Englewood Cliffs, NJ: Prentice-Hall.

Fung, I. Y., S. K. Meyn, I. Tegen *et al.* 2000. Iron supply and demand in the upper ocean. *Global Biogeochemical Cycles*, **14**, 281–295.

Futrelle, R. P. 1984. How molecules get to their detectors. The physics of diffusion of insect pheromones. *Trends in Neuroscience*, **7**, 116–120.

Gambolati, G., G. Ricceri, W. Bertoni, G. Brighenti and E. Vuillermin. 1991. Mathematical simulations of the subsidence of Ravenna. *Water Resources Research*, **27**, 2899–2918.

Gara, R. I., W. R. Littke, J. K. Agee *et al.* 1985. Influence of fires, fungi, and mountain pine beetles on development of a lodgepole pine forest in south-central Oregon. In *Lodgepole Pine: The Species and its Management*, ed. D. M. Baumgartner, R. G. Krebill, J. T. Arnott and G. F. Weetman, pp. 153–162. Pullman, WA: Washington State University Cooperative Extension Service.

Gardiner, C. W. 1985. *Handbook of Stochastic Methods for Physics, Chemistry, and the Natural Sciences*. Berlin: Springer-Verlag.

Gardner, R. H. and R. V. O'Neill. 1991. Pattern, process, and predictability: the use of neutral models for landscape analysis. In *Quantitative Methods in Landscape Ecology*, ed. M. G. Turner and R. H. Gardner, pp. 289–307. New York: Springer-Verlag.

Gardner, R. H., R. V. O'Neill, M. G. Turner and V. H. Dale. 1989. Quantifying scale-dependent effects of animal movement with simple percolation models. *Landscape Ecology*, **3**, 217–227.

Gardner, R. H., W. H. Romme and M. G. Turner. 1999. Predicting forest fire effects at landscape scales. In *Spatial Modeling of Forest Landscapes: Approaches and Applications*, ed. D. J. Mladenoff and W. L. Baker, pp. 163–185. Cambridge, UK: Cambridge University Press.

Garratt, J. R. 1992. *The Atmospheric Boundary Layer*. Cambridge, UK: Cambridge University Press.

Gerhardt, H. C. 1975. Sound pressure levels and radiation patterns of the vocalizations of some North American frogs and toads. *Journal of Comparative Physiology*, **102**, 1–12.

Germino, M. J., W. A. Reiners, B. J. Blasko, D. McLeod and C. T. Bastion. 2001. Estimating visual properties of Rocky Mountain landscapes using GIS. *Landscape and Urban Planning*, **53**, 71–83.

Gill, F. B. and L. R. Wolf. 1975. Economics of feeding territoriality in the golden-winged sunbird. *Ecology*, **56**, 333–345.

Giorgi, F., M. R. Marinucci, and G. Visconti. 1990. Use of a limited-area model nested in a general circulation model for regional climate simulation over Europe. *Journal of Geophysical Research*, **95**, 18, 413–18, 431.

GIS/EM4 (Geographic Information Systems/Environmental Modeling 4). 2000. *Proceedings of the 4th International Conference on Integrating Geographic Information Systems (GIS) and Environmental Modeling*. CD-ROM.

Gleason, H. A. 1926. The individualist concept of the plant association. *Bulletin of the Torrey Botanical Club*, **53**, 7–26.

Goldsmith, T. H. 1990. Optimization, constraint, and history in the evolution of eyes. *Quarterly Review of Biology*, **65**, 281–322.

Goodchild, M. F. 1992. Geographical information science. *International Journal of Geographical Information Systems*, **6**, 31–45.

Goodchild, M. F., B. O. Parks and L. T. Steyaert (ed.) 1993. *Environmental Modeling with GIS*. New York: Oxford University Press.

Goodchild, M. F., L. T. Steyaert, B. O. Parks *et al.* (ed.) 1996. *GIS and Environmental Modeling: Progress and Research Issues*. Ft. Collins, CO: GIS World Books.

Goodison, B. E., H. L. Ferguson and G. A. McKay. 1981. Measurement and data analysis. In *Handbook of Snow: Principles, Processes, Management and Use*, ed. D. M. Gray and D. H. Male, pp. 191–274. Toronto: Pergamon Press.

Grace, E. S. 1997. *The World of the Monarch Butterfly*. San Francisco, CA: Sierra Club Books.

Graham, J. 1984. Methods of stability analysis. In *Slope Instability*, ed. D. Brundsen and D. B. Prior, pp. 171–215. Chichester, UK: Wiley.

Gray, W. M., J. D. Sheaffer and C. W. Landsea. 1997. Climate trends associated with multidecadal variability of Atlantic hurricane activity. In *Hurricanes: Climate and Socioeconomic Impacts*, ed. H. F. Diaz and R. S. Pulwarty, pp. 15–53. New York: Springer-Verlag.

Green, J. 1999. *Atmospheric Dynamics*. Cambridge, UK: Cambridge University Press.

Greene, B. 1999. *The Elegant Universe: Superstrings, Hidden Dimensions, and the Quest for the Ultimate Theory*. New York: W. W. Norton.

Greene, C. H., P. H. Wiebe and J. E. Zamon. 1994. Acoustic visualization of patch dynamics in oceanic ecosystems. *Oceanography*, **7**, 4–12.

Greene, C. H., P. H. Wiebe, C. Pelkie, M. C. Benfield and J. M. Popp. 1998. Three-dimensional acoustic visualization of zooplankton patchiness. *Deep-Sea Research II*, **45**, 1201–1217.

Greene, D. F. and E. A. Johnson. 1995. Long-distance wind dispersal of tree seeds. *Canadian Journal of Botany*, **73**, 1036–1045.

Greenwood, P. J. and I. R. Swingland. 1983. Animal movement: approaches, adaptations and constraints. In *The Ecology of Animal Movment*, ed. I. R. Swingland and P. J. Greenwood, pp. 1–6. Oxford: Oxford Science Publications, Clarendon Press.

Groffman, P. M., R. Brumme, K. Butterbach-Bahl *et al.* 2000. Evaluating annual nitrous oxide fluxes at the ecosystem scale. *Global Biogeochemical Cycles*, **14**, 1061–1070.

Gruell, G. E. 1985. Fire on the early western landscape: an annotated record of wildland fires 1776–1900. *Northwest Science*, **59**, 97–107.

Hagerstrand, T. 1967. *Innovation Diffusion as a Spatial Process*. Chicago, IL: University of Chicago Press.

Haines-Young, R. 1999. Landscape pattern: context and process. In *Issues in Landscape Ecology*, ed. J. A. Wiens and M. R. Moss, pp. 33–37. Ft. Collins, CO: International Association for Landscape Ecology.

Hairston, N. G., F. E. Smith and L. B. Slobodkin. 1960. Community structure, population control, and competition. *American Naturalist*, **94**, 421–425.

Halvorson, J. J., H. Bolton, Jr and J. L. Smith. 1997. The pattern of soil variables related to *Artemisia tridentata* in a burned shrub–steppe site. *Soil Science Society of America Journal*, **61**, 287–294.

Hansen, A. 1984. Landslide hazard analysis. In *Slope Instability*, ed. D. Brundsen and D. B. Prior, pp. 523–602. Chichester, UK: Wiley.

Hansen, A. J. and F. di Castri. 1992. *Landscape Boundaries. Consequences for Biotic Diversity and Ecological Flows*. New York: Springer-Verlag.

Hanski, I. 1998. Metapopulation dynamics. *Nature*, **396**, 41–49.

1999. *Metapopulation Ecology*. Oxford, UK: Oxford University Press.

Hanski, I. and M. E. Gilpin. 1991. *Metapopulation Dynamics*. San Diego, CA: Academic Press.

Hanson, H. C. 1962. *Dictionary of Ecology*. Washington, DC: Philosophical Library.

Hansson, L., L. Fahrig and G. Merriam. 1995. *Mosaic Landscapes and Ecological Processes*, London: Chapman & Hall.

Harding, J. H. 1997. *Amphibians and Reptiles of the Great Lakes Region*. Ann Arbor, MI: University of Michigan Press.

Hargrove, W. W., R. H. Gardner, M. G. Turner, W. H. Romme, and D. G. Despain. 2000. Simulating fire patterns in heterogeneous landscapes. *Ecological Modelling*, **135**, 243–263.

Harris, C. J. (ed.) 1979. *Proceedings of the Conference on Mathematical Modelling of Turbulent Diffusion in the Environment*, Liverpool, September 1978. New York: Academic Press.

Harris, J. G. and M. W. Harris. 1994. *Plant Identification Terminology. An Illustrated Glossary*. Spring Lake, UT: Spring Lake Publishing.

Hasse, L. and F. Dobson. 1986. *Introductory Physics of the Atmosphere and Ocean*. Dordrecht: Reidel.

Hemond, H. F. and E. J. Fechner. 1994. *Chemical Fate and Transport in the Environment*. San Diego, CA: Academic Press.

Hemond, H. F. and E. J. Fechner-Levy. 2000. *Chemical Fate and Transport in the Environment*, 2nd edn. San Diego, CA: Academic Press.

Hergarten, S. and H. J. Neugebauer. 1996. A physical statistical approach to erosion. *Geologische Rundschau*, **85**, 65–70.

Herring, P. J. 2000. Species abundance, sexual encounter and bioluminescent signaling in the deep sea. *Philosophical Transactions of the Royal Society of London*, **355**, 1273–1276.

Hickey, J. R. 1997. The Dispersal of Seeds of Understory Shrubs by American Martens, *Martes americana*, on Chichagof Island, Alaska. M.Sc. Thesis, Department of Zoology and Physiology, University of Wyoming, Laramie, WY.

Hickey, J. R., R. W. Flynn, S. W. Buskirk, K. G. Gerow and M. F. Wilson. 1999. An evaluation of a mammalian predator, *Martes americana*, as a disperser of seeds. *Oikos*, **87**, 499–508.

Hicks, B. B. and D. R. Matt. 1987. Combining biology, chemistry, and meteorology in modeling and measuring dry deposition. *Journal of Atmospheric Chemistry*, **6**, 117–131.

Hiemstra, C. A. 1999. Wind Redistribution of Snow at Treeline, Medicine Bow Mountains, Wyoming. M.Sc. Thesis, Department of Botany, University of Wyoming, Laramie, WY.

Hiemstra, C. A., G. E. Liston and W. A. Reiners. 2002. Snow redistribution by wind and interactions with vegetation at upper treeline in the Medicine Bow Mountains, Wyoming, USA. *Arctic, Antarctic and Alpine Research*, **34**, 262–273.

Hill, K. G., J. J. Loftus-Hills, and D. F. Gartside. 1972. Pre-mating isolation between the australian field crickets *Teleogryllus commodus* and *T. oceanicus* (Orthoptera: Gryllidae). *Australian Journal of Zoology*, **20**, 153–163.

Hill, M. C. 1990. *MODFLOWP: A Computer Program for Estimating Parameters of a Transient, Three-dimensional Ground-water Flow Model Using Nonlinear Regression. Open File Report 91-484*. Denver, CO: US Geological Survey.

Hillel, D. 1998. *Environmental Soil Physics*. San Diego, CA: Academic Press.

Hobbs, N. T. 1996. Modification of ecosystems by ungulates. *Journal of Wildlife Management*, **60**, 695–713.

Holling, C. S. 1959. The components of predation as revealed by a study of small mammal predation of the European pine sawfly. *Canadian Entomologist*, **91**, 293–525.

Holt, R. D. 1993. Ecology at the mesoscale: the influence of regional processes on local communities. In *Species Diversity in Ecological Communities*, ed. R. E. Ricklefs and D. Schluter, pp. 77–88. Chicago, IL: University of Chicago Press.

Holtmeier, F. and G. Broll. 1992. The influence of tree islands and microtopography on pedoecological conditions in the forest–alpine tundra ecotone on Niwot Ridge, Colorado, Front Range, USA. *Arctic and Alpine Research*, **24**, 216–228.

Hoogvelt, A. M. 1976. *The Sociology of Developing Societies*. London: Macmillan.

Hopp, S. L., M. J. Owren and C. S. Evans (ed.) 1998. *Animal Acoustic Communication: Sound Analysis and Research Methods*. Heidelberg: Springer-Verlag.

Horn, H. S. 1983. Some theories about dispersal. In *The Ecology of Animal Movement*, ed. I. R. Swingland and P. J. Greenwood, pp. 54–62. Oxford: Oxford Science Publications, Clarendon Press.

Horn, H. S. and R. H. MacArthur. 1972. Competition among fugitive species in a harlequin environment. *Ecology*, **53**, 749–752.

Huffaker, C. B. 1958. Experimental studies on predation: dispersion factors and predator–prey oscillations. *Hilgardia*, **27**, 343–383.

Hungerford, R. D., M. G. Harrington, W. H. Fraudsen, K. C. Ryan and G. J. Niehoff. 1991. Influence of fire on factors that affect site productivity. In *Management and Productivity of Western-montane Forest Soils (General Technical Report INT-280)*, A. E. Harvey and L. F. Neuenschwander (compilers), pp. 32–50. Ogden, UT: USDA Forest Service.

Ims, R. A. 1995. Movement patterns related to spatial structures. In *Mosaic Landscapes and Ecological Processes*, ed. L. Hansson, L. Fahrig and G. Merriam, pp. 85–109. London: Chapman & Hall.

IPCC, 2001a. Climate Change 2000. *The Science of Climate Change. Contribution of Working Group I to the Third Assessment Report of the Intergovernmental Panel on Climate Change*. Cambridge: Cambridge University Press.
 2001b. *Intergovernmental Panel on Climate Change. Contributions of the IPCC Working Groups to the Third Assessment Report*. Cambridge: Cambridge University Press.

Isard, S. A. and S. H. Gage. 2001. *Flow of Life in the Atmosphere. An Airscape Approach to Understanding Invasive Organisms*. East Lansing, MI: Michigan State University Press.

Iverson, J. D. 1980. Wind Tunnel Modelling of Snow Fences and Natural Snowdrift Controls. In *Proceedings of the 37th Eastern Snow Conference*, Peterborough, Ontario, pp. 106–124.

Iverson, R. M. 1997. The physics of debris flows. *Reviews of Geophysics*, **35**, 245–296.

Jackson, S. T. and M. E. Lyford. 1999. Pollen dispersal models in Quaternary plant ecology: assumptions, parameters and prescriptions. *Botanical Review*, **65**, 39–75.

Jakeman, A. J., T. R. Green, L. Zhang *et al.* 1998. Modelling catchment erosion, sediment and nutrient transport in large basins. In *Soil Erosion at Multiple Scales. Principles and Methods for Assessing Causes and Impacts*, ed. F. W. T. Penning de Vries, F. Agus and J. Kerr, pp. 343–355. Wallingford, UK: CABI.

Janson, C. H., E. W. Stiles and D. W. White. 1986. Selection on plant fruiting traits by brown capuchin monkeys: a multivariate approach. In *Frugivores and Seed Dispersal*, ed. E. Estrada and T. H. Fleming, pp. 83–92. Dordrecht: W. Junk.

Jarvis, P. G. and K. G. McNaughton. 1986. Stomatal control of transpiration: scaling up from leaf to region. *Advances in Ecological Research*, **15**, 1–49.

Jickells, T. D., S. Dorling, W. G. Deuser *et al.* 1998. Air-borne dust fluxes to a deep water sediment trap in the Sargasso Sea. *Global Biogeochemical Cycles*, **12**, 311–320.

Johnson, A. M. 1996. *A Model for Grain Flow and Debris Flow. Open-File Report 96-728*. Washington, DC: US Geological Survey.

Johnson, C. G. 1969. *Migration and Dispersal of Insects by Flight*. London: Methuen.

Johnson, E. A. 1992. *Fire and Vegetation Dynamics: Studies from the North American Boreal Forest*. Cambridge: Cambridge University Press.

Johnson, K. S. 2001. Iron supply and demand in the upper ocean: is extraterrestrial dust a significant source of bioavailable iron? *Global Biogeochemical Cycles*, **15**, 61–63.

Johnston, C. A. 1998. *Geographic Information Systems in Ecology*. Oxford: Blackwell Science.

Jones, C. G., R. S. Ostfeld, M. P. Richard, E. M. Schauber and J. O. Wolff. 1998. Chain reactions linking acorns to gypsy moth outbreaks and Lyme disease risk. *Science*, **279**, 1023–1025.

Jones, J. A. A. 1997. *Global Hydrology. Processes, Resources and Environmental Management*. Harlow, UK: Addison Wesley Longman.

Jones, R. E. 1977. Movement patterns and egg distribution in cabbage butterflies. *Journal of Animal Ecology*, **46**, 195–212.

Julien, P. Y. 1999. *Erosion and Sedimentation*. Cambridge, UK: Cambridge University Press.

Kabailiene, M. V. 1969. On formation of pollen spectra and restoration of vegetation. *Transactions of the Institute of Geology, Vilnius*, **11**, 1–148.

Kahl, J. D. W. 1996. On the prediction of trajectory model error. *Atmospheric Environment*, **30**, 2945–2957.

Kahl, J. D. W., R. C. Schnell, P. J. Sheridan *et al*. 1991. Predicting atmospheric debris transport in real-time using a trajectory forecast model. *Atmospheric Environment*, **25A**, 1705–1713.

Kamil, A. C., J. R. Krebs and H. R. Pulliam. 1987. *Foraging Behavior*. New York: Plenum Press.

Kangas, P. C. 1989. An energy theory of landscape for classifying wetlands. In *Forested Wetlands of the World*, ed. A. Lugo, M. Brinson, and S. Brown, pp. 15–23, Amsterdam: Elsevier.

Kantha, L. H. and C. A. Clayson. 2000a. *Numerical Models of Oceans and Oceanic Processes*. San Diego, CA: Academic Press.
2000b. *Small Scale Processes in Geophysical Fluid Flows*. San Diego, CA: Academic Press.

Kareiva, P. 1982. Experimental and mathematical analysis of herbivore movement: quantifying the influence of plant spacing and quality on foraging discrimination. *Ecological Monographs*, **52**, 261–282.

Kareiva, P. and U. Wennergren. 1995. Connecting landscape patterns to ecosystem and population processes. *Nature*, **373**, 299–301.

Kawashima, S. and Y. Takahashi. 1995. Modelling and simulation of mesoscale dispersion processes for airborne cedar pollen. *Grana*, **34**, 142–150.

Keane, R. E., P. Morgan and S. W. Running. 1996. *Fire-BGC: A Mechanistic Ecological Process Model for Simulating Fire Succession on Coniferous Forest Landscapes of the Northern Rocky Mountains. Research Paper INT-RP-484*. Ogden, UT: USDA Forest Service Intermountain Research Station.

Keane, R. E., C. C. Hardy, K. C. Ryan and M. A. Finney. 1997. Simulating effects of fire on gaseous emissions from future landscapes of Glacier National Park, Montana, USA. *World Resource Review*, **9**, 177–205.

Keitt, T. H. 2000. Spectral representation of neutral landscapes. *Landscape Ecology*, **15**, 479–493.

Keller, M. and W. A. Reiners. 1994. Soil–atmosphere exchange of nitrous oxide and methane under secondary succession of pasture to forest in the Atlantic lowlands of Costa Rica. *Global Biogeochemical Cycles*, **8**, 399–409.

Kelmelis, J. A. 1998. Process dynamics, temporal extent, and causal propagation as the basis for linking space and time. In *Spatial and Temporal Reasoning in Geographic Information Systems*, ed. M. J. Egenhofer and R. G. Golledge, pp. 94–103. New York: Oxford University Press.

Kessler, A. and I. T. Baldwin. 2001. Defensive function of herbivore-induced plant volatile emissions in nature. *Science*, **291**, 2141–2144.

Kimura, M. 1982. Changes in population structure, productivity and dry matter allocation with progress of wave regeneration of *Abies* stands in Japanese subalpine regions. In *Carbon Uptake and Allocation in Subalpine Ecosystems as a Key to Management*, ed. R. H. Waring, pp. 57–63. Corvallis, OR: Oregon State University Press.

Kind, R. J. 1981. Snow drifting. In *Handbook of Snow: Principles, Processes, Management and Use*, ed. D. M. Gray and D. H. Male, pp. 338–359. Toronto: Pergamon Press.
1986. Snowdrifting: a review of modelling methods. *Cold Science Research Technology*, **12**, 217–228.

Kittel, T. G. F., N. A. Rosenbloom, T. H. Painter and VEMAP Modeling Participants. 1995. The VEMAP integrated database for modeling United States ecosystem/ vegetation sensitivity to climate change. *Journal of Biogeography*, **22**, 857–862.

Kittle, J. L., Jr, A. M. Lumb, P. R. Hummel, P. B. Duda and M. H. Gray. 1998. *A Tool for the Generation and Analysis of Model Simulation Scenarios for Watersheds (GenScn). Water-Resources Investigations Report 98-4134* Denver, CO: US Geological Survey.

Knight, D. H. 1994. *Mountains and Plains. The Ecology of Wyoming Landscapes*. New Haven, CT: Yale University Press.

Knisel, W. 1980. *CREAMS: A Field Scale Model for Chemicals, Runoff, and Erosion from Agricultural Management Systems*. Conservation Research Report No. 26. Washington, DC: US Department of Agriculture.

Kolasa, J. and C. D. Rollo. 1991. Introduction: the heterogeneity of heterogeneity – a glossary. In: *Ecological Heterogeneity*, eds. J. Kolasa and S. T. A. Pickett, pp. 1–23. New York: Springer-Verlag.

Kolata, G. 1984. Managing the inland sea: the last large tract of tallgrass prairie in Kansas where ecologists are attempting to maintain it as it was in presettlement days. *Science*, **224**, 703–704.

Kooser, J. G. and R. E. J. Boerner. 1985. Rates and patterns of leaf litter redistribution within an unglaciated Ohio watershed. *Ohio Journal of Science*, **85**, 90–91.

Korb, J. and K. E. Linsenmair. 1999. The architecture of termite mounds: a result of a trade-off between thermoregulation and gas exchange? *Behavioral Ecology*, **10**, 312–316.

Koster, R. D. and M. J. Suarez. 1992. Modeling the land surface boundary in climate models as a composite of independent vegetation stands. *Journal of Geophysical Research*, **97**, 2697–2715.

Kot, M., M. A. Lewis and P. van den Driessche. 1996. Dispersal data and the spread of invading organisms. *Ecology*, **77**, 2027–2042.

Kowalik, W. S., S. E. Marsh and R. J. P. Lyon. 1982. A relation between Landsat digital numbers, surface reflectance, and the cosine of the solar zenith angle. *Remote Sensing of Environment*, **12**, 39–55.

Kustas, W. P., B. J. Choudhury, K. E. Kunkel and L. D. Gay. 1989. Estimate of the aerodynamic roughness parameters over an incomplete canopy cover of cotton. *Agricultural and Forest Meteorology*, **46**, 91–105.

Laflen, J. M., L. J. Lane and G. R. Foster. 1991. WEPP. A new generation of erosion prediction technology. *Journal of Soil and Water Conservation*, **46**, 34–38.

Lai, S. and P. K. Patra. 1998. Variabilities in the fluxes and annual emissions of nitrous oxide from the Arabian Sea. *Global Biogeochemical Cycles*, **12**, 321–327.

Lane, L. J. and M. A. Nearing 1989. *USDA Water Erosion Prediction Project: Hillslope Profile Model Documentation*. NSERL Report No. 2. West Lafayette, IN: USDA-ARS National Soil Erosion Research Laboratory.

Lao-tzu, translated by Lau, D. C. 1963. *Tao Te Ching*. London: Penguin Classics.

Larcher, W. 1995. *Physiological Plant Ecology. Ecophysiology and Stress Physiology of Functional Groups*, 3rd edn. Berlin: Springer-Verlag.

Larom, D., M. Garstang, M. Lindeque *et al.* 1997a. Meteorology and elephant infrasound at Etosha National Park, Namibia. *Journal of the Acoustical Society of America*, **101**, 1710–1717.

Larom, D., M. Garstang, K. Payne, R. Raspet and M. Lindeque. 1997b. The influence of surface atmospheric conditions on the range and area reached by animal vocalizations. *Journal of Experimental Biology*, **200**, 421–431.

Latham, D. J. and R. C. Rothermel. 1993. *Probability of fire-stopping precipitation events*. Research Note INT-410. Ogden, UT: USDA Forest Service, Intermountain Research Station.

Laursen, S. 2001. Presentation, Release, Transport and Deposition of *Artemisia Tridentata* Nutt. Pollen in a Sagebrush Steppe. M.Sc. Thesis. Department of Botany, University of Wyoming, Laramie, WY.

Laws, E. A., P. G. Falkowski, W. O. Smith, Jr, D. Ducklow and J. J. McCarthy. 2000. Temperature effects on export production in the open ocean. *Global Biogeochemical Cycles*, **14**, 1231–1246.

Leavesley, G. H., R. W. Lichty, B. M. Troutman and L. G. Saindon. 1983. *Precipitation-Runoff Modeling System: User's Manual: Water-Resources Investigations Report 83-4238*. Denver, CO: US Geological Survey.

Lemonick, M. D. 2001. Life in the greenhouse. *Time Magazine*, **157**, 9 April.

Lenton, T. M. and A. J. Watson. 2000. Redfield revisited 1. Regulation of nitrate, phosphate, and oxygen in the ocean. *Global Biogeochemical Cycles*, **14**, 225–248.

Leopold, A. 1941. Lakes in relation to terrestrial life patterns. In *A Symposium on Hydrobiology* [ed. none listed], pp. 17–22. Madison, WI: University of Wisconsin Press.

Lepart, J. and M. Debussche. 1992. Human impacts on landscape patterning: Mediterranean examples. In *Landscape Boundaries: Ecological Consequences on Biotic Diversity and Ecological Flows*. ed. A. J. Hansen and F. di Castri, pp. 76–106. New York: Springer-Verlag.

Levin, S. A. and R. T. Paine. 1974. Disturbance, patch formation, and community structure. *Proceedings of the National Academy of Sciences USA*, **71**, 2744–2747.

Levine, J. S. (ed.) 1991. *Global Biomass Burning: Atmospheric, Climatic and Biospheric Implications*. Cambridge, MA: MIT Press.

Levins, R. 1968. *Evolution in Changing Environments*. Princeton, NJ: Princeton University Press.

1970. Extinction. *Lecture Notes in Mathematics*, **2**, 75–107.

Levinton, J. S. 1995. *Marine Biology: Function, Diversity, Ecology*. New York: Oxford University Press.

Li, R. 2000. Data models for marine and coastal geographic information systems. In *Marine and Coastal Geographical Information Systems*, ed. D. J. Wright, and D. J. Bartlett, pp. 25–36. London: Taylor & Francis.

Lidicker, W. Z., Jr and R. L. Caldwell (ed.) 1982. *Dispersal and Migration*. Stroudsburg, PA: Hutchinson Ross.

Liebenow, A. M., W. J. Elliot, J. M. Laflen and K. D. Kohl. 1990. Interrill erodability: collection and analysis of data from cropland soils. *Transaction of the ASAE*, **33**, 1882–1888.

Lindeboom, H. L. 1984. The nitrogen pathway in a penguin rookery. *Ecology*, **65**, 269–277.

Lindeman, R. L. 1942. The trophic–dynamic aspect of ecology. *Ecology*. **23**, 399–418.

Liston, G. E. 1999. Interrelationships between snow distribution, snowmelt, and snowcover depletion: implications for atmospheric, hydrologic, and ecologic modeling. *Journal of Applied Meteorology*, **36**, 205–213.

Liston, G. E. and M. Sturm. 1998. A snow-transport model for complex terrain. *Journal of Glaciology*, **44**, 498–516.

Liu, S., W. A. Reiners, M. Keller and D. S. Schimel. 1999. Model simulation of changes in N_2O and NO emissions with conversion of tropical rain forests to pastures in the Costa Rican Atlantic Zone. *Global Biogeochemical Cycles*, **13**, 663–677.

Livingston, G. P. and G. L. Hutchinson. 1995. Enclosure-based measurement of trace gas exchange: applications and sources of error. In *Biogenic Trace Gases: Measuring Emissions from Soil and Water*, ed. P. A. Matson and R. C. Harriss, pp. 14–51. Cambridge, UK: Blackwell Science.

Lloyd, J. E. 1966. *Studies on the Flash Communication System in* Photinus *Fireflies*. *Miscellaneous Publication No. 130*. Ann Arbor, MI: University of Michigan Museum of Zoology.

Longley, P. A., M. F. Goodchild, D. J. Maguire and D. W. Rhind, (ed.) 1999. *Geographical Information Systems. Principles and Technical Issues*, 2nd edn. New York: Wiley.

2001. *Geographic Information Systems and Science*. Chichester, UK: Wiley.

Lotka, A. J. 1925. *Elements of Physical Biology*. Baltimore, MD: Williams & Wilkins.

Lovett, G. M. and W. A. Reiners. 1986. Canopy structure and cloud water deposition in subalpine coniferous forests. *Tellus*, **38B**, 319–327.

Lucas, L. V., J. R. Koseff, S. G. Monismith, J. E. Cloern and J. K. Thompson. 1999. Processes governing phytoplankton blooms in estuaries. II. The role of horizontal transport. *Marine Ecology Progress Series*, **187**, 17–30.

Luckman, B. H. 1971. The role of snow avalanches in the evolution of alpine talus slopes. In *Slopes: Form and Process*, ed. D. Brundsen, pp. 93–110. *Special Publication No. 3*. London: Institute of British Geographers.

Ludwig, D., D. D. Jones and C. S. Holling. 1978. Qualitative analysis of insect outbreak systems: spruce budworm and forest. *Journal of Animal Ecology*, **47**, 315–332.

Ludwig, J. A., D. Tongway, D. Freudenberger, J. Noble and K. Hodgkinson (ed.) 1997.

Landscape Ecology Function and Management. Principles from Australia's Rangelands. Canberra: CSIRO Publishing.

Ludwig, J. A., J. A. Wiens, and D. J. Tongway. 2000. A scaling rule for landscape patches and how it applies to conserving soil resources in savannas. *Ecosystems*, **3**, 84–97.

MacArthur, R. H. 1958. Population ecology of some warblers of northeastern coniferous forests. *Ecology*, **39**, 599–619.

1967. The limiting similarity, convergence, and divergence of coexisting species. *American Naturalist*, **101**, 377–385.

MacArthur, R. H. and R. Levins. 1964. Competition, habitat selection, and character displacement in a patchy environment. *Proceedings of the National Academy of Sciences USA*, **51**, 1207–1210.

MacArthur, R. H. and E. R. Pianka. 1966. On the optimal use of a patchy environment. *American Naturalist*, **100**, 603–609.

MacArthur, R. H. and E. O. Wilson. 1967. *The Theory of Island Biogeography*. Princeton, NJ: Princeton University Press.

MacKinnon, J. R. 1974. The ecology and behaviour of wild orangutans (*Pongo pygmaeus*). *Animal Behavior*, **22**, 3–74.

Maclean, N. 1976. *A River Runs Through It, and Other Stories*. Chicago, IL: University of Chicago Press.

1992. *Young Men and Fire*. Chicago, IL: University of Chicago Press.

Maguire, D. J. and J. Dangermond. (ed.) 1991. The functionality of GIS. In *Geographical Information Systems*, Vol. 1, ed. D. J. Maguire, M. F. Goodchild and D. W. Rhind, pp. 319–325. New York: Wiley.

Maguire, D. J., M. F. Goodchild and D. W. Rhind (ed.) 1991. *Geographical Information Systems. Principles and Applications*. New York: Wiley.

Malanson, G. P. 1995. *Riparian Landscapes*. Cambridge, UK: Cambridge University Press.

Margalef, R. 1963. On certain unifying principles in ecology. *American Naturalist*, **97**, 357–374.

Mark, D. M. 1999. Spatial representation: a cognitive view. In *Geographical Information Systems*, 2nd edn., ed. P. Longley, M. F. Goodchild, D. J. Maguire and D. W. Rhind, pp. 81–89. New York: Wiley.

Marshall, N. J. 2000. Communication and camouflage with the same 'bright' colours in reef fishes. *Philosophical Transactions of the Royal Society of London*, **355**, 1243–1248.

Marten, K. and P. Marler. 1977. Sound transmission and its significance for animal vocalization. *Behavioral Ecology and Sociobiology*, **2**, 271–290.

Martin, D. J. 1999. Spatial representation: the social scientist's perspective. In *Geographical Information Systems*, 2nd edn., ed. P. Longley, M. F. Goodchild, D. J. Maguire and D. W. Rhind, pp. 71–80. New York: Wiley.

Massman, W., R. Sommerfeld, K. Zeller, T. Hehn, L. Hudnell and S. Rochelle. 1995. CO_2 flux through a Wyoming seasonal snow pack: diffusional and pressure pumping effects. In *Biogeochemistry of Seasonally Snow-covered Catchments* (*Publication No. 28*), ed. K. A. Tonnessen, M. W. Williams and M. Tranter, pp. 71–79. Wallingford, UK: IAHS Press, Institute of Hydrology.

Massman, W., R. A. Sommerfeld, A. R. Mosier, K. F. Zeller, T. J. Hehn and S. G. Rochelle. 1997. A model investigation of turbulence-driven pressure pumping effects on the rate of diffusion of CO_2, N_2O and CH_4 through layered snow packs. *Journal of Geophysical Research*, **102**, 18851–18863.

Mast, M. A., K. P. Wickland, R. T. Striegl and D. W. Clow. 1998. Winter fluxes of CO_2 and CH_4 from subalpine soils in Rocky Mountain National Park, Colorado. *Global Biogeochemical Cycles*, **12**, 607–620.

Matson, P. A. and R. H. Waring. 1984. Effects of nutrient and light limitation on mountain hemlock: susceptibility to laminated root rot. *Ecology*, **65**, 1517–1524.

Matson, P. A., W. H. McDowell, A. R. Townsend and P. M. Vitousek. 1999. The globalization of nitrogen deposition: ecosystem consequences in tropical environments. *Biogeochemistry*, **46**, 67–83.

McAtee, W. L. 1917. Showers of organic matter. *Monthly Weather Review*, **45**, 217–224.

McCleave, J. D., G. P. Arnold, J. J. Dodson and W. H. Neill. 1984. *Mechanisms of Migration in Fishes*. New York: Plenum Press.

McClung, D. and P. Schaerer. 1993. *The Avalanche Handbook*. Seattle, WA: The Mountaineers.

McDermott, F. A. 1958. *The Fireflies of Delaware*. Wilmington, DE: Society on Natural History.

McDonald, M. G. and A. W. Harbaugh. 1988. *A Modular Three-dimensional Finite-difference Ground-water Flow Model. Techniques of Water-Resources Investigations O6-A1*. Denver, CO: US Geological Survey.

McDowell, W. H., C. Gines Sanchez, C. E. Asbury and C. R. Ramos Perez. 1990. Influence of sea salt aerosols and long range transport on precipitation chemistry at El Verde, Puerto Rico. *Atmospheric Environment*, **24**A, 2813–2821.

McGregor, P. K., Dabelsteen, T., Clark *et al.* 1997. Accuracy of a passive acoustic location system: empirical studies in terrestrial habitats. *Ethology Ecology and Evolution*, **9**, 269–286.

McIntosh, R. P. 1985. *The Background of Ecology: Concept and Theory*. Cambridge, UK: Cambridge University Press.

McLauglin, J. F. and J. Roughgarden. 1993. Species interactions in space. In *Species Diversity in Ecological Communities*, ed. R. E. Ricklefs and D. Schluter, pp. 89–98. Chicago, IL: University of Chicago Press.

McNaughton, S. J., R. W. Ruess and S. W. Seagle. 1988. Large mammals and process dynamics in African ecosystems. *BioScience*, **38**, 794–800.

McNaughton, S. J., F. F. Banyikwa and M. M. McNaughton. 1997. Promotion of the cycling of diet-enhancing nutrients by African grazers. *Science*, **278**, 1798–1800.

Meixner, F. X. and W. Eugster. 1999. Effects of landscape pattern and topography on emissions and transport. In *Integrating Hydrology, Ecosystem Dynamics, and Biogeochemistry in Complex Landscapes*, ed. J. D. Tenhunen and P. Kabat, pp. 147–175. Chichester, UK: Wiley.

Messier, C. and P. Bellefleur. 1988. Light quantity and quality on the forest floor of pioneer and climax stages in a birch–beech–sugar maple stand. *Canadian Journal of Forest Research*, **18**, 615–622.

Meyer, J. L. and E. T. Schultz. 1985. Migrating haemulid fishes as a source of nutrients and organic matter on coral reefs. *Limnology and Oceanography*, **30**, 146–156.

Meyer, L. D. and W. H. Wischmeier. 1969. Mathematical simulation of the process of soil erosion by water. *Transactions ASAE*, **12**, 754–758, 762.

Miller, J. M. 1987. *The Use of Back-air Trajectories in Interpreting Atmospheric Chemistry Data: A Review and Bibliography. NOAA Technical Memo ERL ARL-1550*. Silver Spring, MD: Air Resources Laboratory.

Mitas, L. and H. Mitasova. 1998a. Distributed erosion modeling for effective erosion. *Water Resources Research*, **34**, 505–516.

1998b. Multi-scale Green's function Monte Carlo approach to erosion modelling and its application to land-use optimization. In *Modelling Soil Erosion, Sediment Transport and Closely Related Hydrological Processes*, ed. W. Summer, E. Klaghofer and W. Zhang, pp. 81–90. Wallingford, UK: International Association of Hydrological Sciences.

Mitas, L., H. Mitasova, W. M. Brown and M. Astley. 1996. Interacting fields approach for evolving spatial phenomena; application to erosion simulation for optimized land use. In *Proceedings of the III International Conference on Integration of Environmental Modeling and GIS*, Santa Barbara, CA: National Center for Geographic Information and Analysis [CD-ROM].

Mitasova, H. and L. Mitas. 1998. Unit 130. Process modeling and simulations. In *NCGIA Core Curriculum in Geographic Information Science*. Santa Barbara, CA: National Center for Geographic Information and Analysis.

2002. Modeling physical systems. In *Geographic Information Systems and Environmental Modeling*, ed. B. Parks, M. Crane and K. Clarke, pp. 189–210. Upper Saddle River, NJ: Prentice Hall.

Moffett, M. W. 2000. What's "up"? A critical look at the basic terms of canopy biology. *Biotropica*, **32**, 589–596.

Monteith, J. L. and M. Unsworth. 1995. *Principles of Environmental Physics*, 2nd edn. London: Arnold.

Montgomery, D. R. and W. E. Dietrich. 1994. A physically based model for the topographic control on shallow landsliding. *Water Resources Research*, **30**, 1153–1171.

Mooney, H. A., T. M. Bonnicksen, N. L. Christensen, J. E. Lotan and W. A. Reiners (ed.) 1981. *Proceedings of the Conference, Fire Regimes and Ecosystem Properties. General Technical Report WO-26* Ogden, UT: USDA Forest Service.

Moore, I. D. and G. J. Burch. 1986a. Modelling erosion and deposition: topographic effects. *Transactions ASAE*, **29**, 1624–1630.

1986b. Sediment transport capacity of sheet and rill flow: application of unit stream power theory. *Water Resources Research*, **22**, 1350–1360.

Moore, I. D., R. B. Grayson and A. R. Ladson. 1991. Digital terrain modeling: a review of hydrological, geomorphological and biological applications. *Hydrological Processes*, **5**, 3–30.

Moss, M. R. 1999. Fostering academic and institutional activities in landscape ecology. In *Issues in Landscape Ecology*, ed. J. A. Wiens and M. R. Moss, pp. 138–144. Ft Collins, CO: International Association of Landscape Ecology.

Mowrer, H. T. 1996. *Decision Support Systems for Ecosystem Management: An Evaluation of Existing Systems. General Technical Report RM-GTR-296.* Ft. Collins, CO: USDA Forest Service.

Murnane, J., C. Barton, E. Collins *et al.* 2000. Model estimates hurricane wind speed probabilities. *EOS, Transactions*, **81**, 433–438.

Murray, J. W. 1992. The oceans. In *Global Biogeochemical Cycles*, ed. S. S. Butcher, R. J. Charleson, G. H. Orians, and G. V. Wolfe, pp. 175–211. San Diego, CA: Academic Press.

Murray, J. W., R. T. Barber, M. R. Roman, M. P. Bacon and R. A. Feely. 1994. Physical and biological controls on carbon cycling in the equatorial Pacific. *Science*, 266, 58–65.

Naimen, R. J. and R. E. Bilby (ed.) 1998. *River Ecology and Management. Lessons from the Pacific Coastal Ecoregion.* New York: Springer-Verlag.

NCGIA, 1996. *NCGIA Third International Conference/Workshop on Integrating GIS and Environmental Modeling.* January, Santa Fe, NM. Santa Barbara, CA: National Center for Geographic Information and Analysis. [CD-ROM].

Nearing, M. A. 1997. A single, continuous function for slope steepness influence on soil loss. *Soil Science Society of America Journal*, **61**, 917–919.

Neshyba, S. 1987. *Oceanography. Perspectives on a Fluid Earth.* New York: Wiley.

Neufeldt, V. and D. B. Guralnik. 1994. *Webster's New Universal Unabridged Dictionary.* Avenel, NJ: Barnes & Noble.

Newman, E. A. and P. H. Hartline. 1982. The infrared "vision" of snakes. *Scientific American*, **246**, 116–127.

Nilsson, L. A. 1988. The evolution of flowers with deep corolla tubes. *Nature*, **334**, 147–149.

Noble, I. R. 1999. Effect of landscape framentation, disturbance, and succession on ecosystem function. In *Integrating Hydrology, Ecosystem Dynamics, and Biogeochemistry in Complex Landscapes*, ed. J. D. Tenhunen and P. Kabat, pp. 297–312. Chichester, UK: Wiley.

Nowak, R. M. 1991. *Walker's Mammals of the World*, 5th edn. Baltimore, MD: Johns Hopkins University Press.

Noy-Meir, I. 1995. Interactive effects of fire and grazing on structure and diversity of Mediterranean grasslands. *Journal of Vegetation Science*, **6**, 701–710.

Nye, P. H. and P. B. Tinker. 1977. *Solute Movement in the Soil–Root System.* Berkeley, CA: University of California Press.

O'Connor, D. J. 1988. Models of sorptive toxic substances in freshwater systems I: basic equations. *Journal of Environmental Engineering*, **114**, 507–532.

Odum, E. P. 1953. *Fundamentals of Ecology.* Philadelphia, PA: Saunders.

1959. *Fundamentals of Ecology*, 2nd edn. Philadelphia, PA: Saunders.

Odum, H. T. 1971. *Environment, Power, and Society.* New York: Wiley-Interscience.

Oechel, W. C., G. Vourlitis and S. J. Hastings. 1997. Cold season CO_2 emission from arctic soils. *Global Biogeochemical Cycles*, **11**, 163–172.

Ogle, S. M. 2000. Impacts of Annual Brome Grasses, *Bromus japonicus* and *Bromus tectorum* on the Structure and Function of a Mixed Grass Prairie Ecosystem. Ph. D.

Thesis, Department of Botany, University of Wyoming, Laramie, WY.

Okubo, A. 1980. *Diffusion and Ecological Problems: Mathematical Models.* Heidelberg: Springer-Verlag.

Okubo, A. and S. A. Levin. 2001. *Diffusion and Ecological Problems. Modern Perspectives,* 2nd edn. New York: Springer-Verlag.

Oleson, K. W., K. L. Driese, J. A. Maslanik, W. J. Emery and W. A. Reiners. 1997. The sensitivity of a land surface parameterization scheme to the choice of remotely sensed land-cover datasets. *Monthly Weather Review,* **125,** 1537–1555.

Omernik, J. M. 1987. Ecoregions of the conterminous United States. *Annals of the Association of American Geographers,* **77,** 118–125.

O'Neill, R. V. and A. W. King. 1998. Homage to St Michael: or, why are there so many books on scale? In *Ecological Scale. Theory and Application,* ed. D. L. Peterson and V. T. Parker, pp. 3–15. New York: Columbia University Press.

O'Neill, T. A. and G. W. Witmer. 1991. Assessing the cumulative impacts to elk and mule deer in the Salmon River Basin, Idaho. *Applied Animal Behavior Science,* **29,** 225–239.

Ono, M., T. Igarashi, E. Ohno and M. Sasaki. 1995. Unusual thermal defense by a honeybee against mass attack by hornets. *Nature,* **377,** 334–336.

Openshaw, S. 1991. Developing appropriate spatial analysis methods for GIS. In *Geographical Information Systems. Principles and Applications,* ed. D. F. Maguire, M. F. Goodchild and D. W. Rhind, pp. 389–402. New York: Wiley.

Orndorff, K. A. and G. E. Lang. 1981. Leaf litter redistribution in a West Virginia hardwood forest. *Journal of Ecology,* **69,** 225–235.

Pace, M. L., J. J. Cole and S. R. Carpenter. 1999. Trophic cascades revealed in diverse ecosystems. *Trends in Ecology and Evolution,* **14,** 483–488.

Paine, R. T. 1966. Food web complexity and species diversity. *American Naturalist,* **100,** 65–75.

Paine, R. T. and S. A. Levin. 1981. Intertidal landscapes: disturbance and the dynamics of pattern. *Ecological Monographs,* **51,** 145–178.

Park, S. and D. Wagner. 1997. Incorporating cellular automata simulators as analytical engines in GIS. *Transactions in GIS,* **2,** 213–231.

Parris, K. M. 2002. More bang for your buck: the effect of caller position, habitat and chorus noise on the efficiency of calling in the spring peeper. *Ecological Modelling,* **156,** 213–224.

Parton, W. J., D. S. Schimel, C. V. Cole and D. S. Ojima. 1987. Analysis of factors controlling soil organic matter levels in Great Plains grasslands. *Soil Science Society of America Journal,* **51,** 1173–1179.

Parton, W. J., D. S. Ojima, C. V. Cole and D. S. Schimel. 1994. A general model for soil organic matter dynamics: sensitivity to litter chemistry, texture and management. In *Quantitative Modeling of Soil Forming Processes,* pp. 147–167. *Special Publication 39.* Madison, WI: Soil Science Society of America.

Pastor, J., B. Dewey, R. Moen, M. White, D. Mladenoff and Y. Cohen. 1998. Spatial patterns in the moose–forest–soil ecosystem on Isle Royale, Michigan, USA. *Ecological Applications,* **8,** 411–424.

Paul, R. C. and T. J. Walker. 1979. Arboreal singing in a burrowing cricket, *Anurogryllus arboreus. Journal of Comparative Physiology,* **132,** 217–223.

Pech, R. P. and A. W. Davis. 1987. Reflectance modeling of semiarid woodlands. *Remote Sensing of Environment,* **23,** 365–377.

Pedgley, D. E. 1990. Concentration of flying insects by the wind. *Philosophical Transactions of the Royal Society of London, Series B,* **328,** 631–653.

Peitersen, M. N. 1994. Dynamics of Large-scale Dry-transport Landslides. M.Sc. Thesis, Department of Geology and Geophysics, University of Wyoming, Laramie, WY.

Penning de Vries, F. W. T., F. Agus and J. Kerr. 1998. *Soil Erosion at Multiple Scales. Principles and Methods for Assessing Causes and Impacts.* Wallingford, UK: CABI.

Peterson, D. L. and V. T. Parker. 1998. *Ecological Scale. Theory and Applications.* New York: Columbia University Press.

Peuquet, D. J. 1999. Time in GIS and geographical databases. In *Geographical Information Systems*, 2nd edn, ed. P. Longley, M. F. Goodchild, D. J. Maguire and D. W. Rhind, pp. 91–103. New York: Wiley.

Philip, J. R. and D. A. de Vries. 1957. Moisture movement in porous materials under temperature gradients. *Transactions of the American Geophysical Union*, **38**, 222–232, 594.

Pickett, S. T. A. and M. L. Cadenasso. 1995. Landscape ecology: spatial heterogeneity in ecological systems. *Science*, **269**, 331–334.
      2002. The ecosystem as a multidimensional concept: meaning, model, and metaphor. *Ecosystems*, **5**, 1–10.

Pickett, S. T. A. and P. S. White. 1985. *The Ecology of Natural Disturbance and Patch Dynamics*. San Diego, CA: Academic Press.

Pickles, J. 1999. Arguments, debates, and dialogues: the GIS–social theory debate and the concern for alternatives. In *Geographical Information Systems*, 2nd edn., ed. P. Longley, M. F. Goodchild, D. J. Maguire and D. W. Rhind, pp. 49–60. New York: Wiley.

Pickup, G., R. J. Higgins and I. Grant. 1983. Modeling sediment transport as a moving wave: the transfer and deposition of mining waste. *Journal of Hydrology*, **60**, 281–301.

Pielke, R. A., W. R. Cotton, R. L. Walko *et al.* 1992. A comprehensive meteorological modeling system: RAMS. *Meteorology and Atmospheric Physics*, **49**, 69–91.

Pielke, R. A., J. Baron, T. Chase *et at.* 1996. Use of mesoscale models for simulation of seasonal weather and climate change for the Rocky Mountain states. In *GIS and Environmental Modeling: Progress and Research Issues*, ed. M. F. Goodchild, L. T. Steyaert, B. O. Parks *et al.*, pp. 99–103. Ft Collins, CO: GIS World Books.

Pierce, G. J. and J. G. Ollason. 1987. Eight reasons why optimal foraging theory is a complete waste of time. *Oikos*, **49**, 111–118.

Pinkel, R., W. Munk, P. Worcester *et al.* 2000. Ocean mixing studied near Hawaiian Ridge. *EOS*, **81**, 545, 553.

Plant, R. A. J. 2000. Regional analysis of soil–atmosphere nitrous oxide emissions in the Northern Atlantic Zone of Costa Rica. *Global Change Biology*, **6**, 639–653.

Poland, J. F. 1973. Subsidence in the United States due to ground-water overdraft: a review. In *Irrigation and Drainage Division Specialty Conference on Agricultural and Urban Considerations in Irrigation and Drainage*, Fort Collins, CO. April 1973.

Polis, G. A., W. B. Anderson and R. D. Holt. 1997. Toward an integration of landscape and food web ecology: the dynamics of spatially subsidized food webs. *Annual Review of Ecology and Systematics*, **28**, 289–316.

Pomeroy, J. W., D. M. Gray and P. G. Landine. 1993. The prairie blowing snow model: characteristics, validation, operation. *Journal of Hydrology*, **144**, 165–192.

Poole, J. H. 1999. Signals and assessment in African elephants: evidence from playback experiments. *Animal Behavior*, **58**, 185–193.

Potter, C. S., P. A. Matson, P. M. Vitousek and E. A. Davidson. 1996. Process modeling of controls on nitrogen trace gas emission from soils worldwide. *Journal of Geophysical Research*, **101**, 1361–1377.

Prentice, I. C. 1985. Pollen representation, source area, and basin size: toward a unified theory of pollen analysis. *Quaternary Research*, **23**, 76–86.
      1988. Records of vegetation in space and time: the principles of pollen analysis. In *Vegetation History*, ed. B. Huntley and T. Webb III, pp. 17–42. Dordrecht: Kluwer Academic.

Preston, E. M. and B. L. Bedford. 1988. Evaluating cumulative effects on wetland functions: a conceptual overview and generic framework. *Environmental Management*, **125**, 565–583.

Prestwich, K. N. 1994. The energetics of acoustic signaling in anurans and insects. *American Zoologist*, **34**, 625–643.

Prospero, J. M., R. A. Glaccum and R. T. Nees. 1981. Atmospheric transport of soil dust from Africa to South America. *Nature*, **289**, 570–572.

Pyke, G. H. 1983. Animal movements: an optimal foraging approach. In *The Ecology of Animal Movement*, ed. I. R. Singland and

P. J. Greenwood, pp. 7–31. Oxford: Oxford Science Publications, Clarendon Press.

Pyne, S. J. 1982. *Fire in America. A Cultural History of Wildland and Rural Fire*. Princeton, NJ: Princeton University Press.

1995. *World Fire: The Culture of Fire on Earth*. New York: Henry Holt.

Pyne, S. J., P. L. Andrews and R. D. Laven. 1996. *Introduction to Wildland Fire*, 2nd edn. New York: Wiley.

Quenzer, A. M. 1998. *A GIS Assessment of the Total Loads and Water Quality in the Corpus Christi Bay System*. CRWR Online Report 98-1. *http://www.ce.utexas.edu/centers/crwr/reports/online.html*

Raper, J. F. (ed.) 1989. *Three-dimensional Applications in Geographical Information Systems*. London: Taylor and Francis.

Raper, J. F. 1999. Spatial representation: the scientist's perspective. In *Geographical Information Systems*, 2nd edn., ed. P. Longley, M. F. Goodchild, D. J. Maguire and D. W. Rhind, pp. 61–70. New York: Wiley.

Raper, J. F. and B. Kelk. 1991. Three-dimensional GIS. In *Geographical Information Systems. Principles and Applications*, ed. D. J. Maguire, M. F. Goodchild and D. W. Rhind, pp. 299–317. New York: Wiley.

Raupach, M. R. 1995. Vegetation–atmosphere interaction and surface conductance at leaf, canopy and regional scales. *Agricultural and Forest Meteorology*, **73**, 151–179.

Raupach, M. R., R. A. Antonia, and S. Rajagopalan. 1991. Rough-wall turbulent boundary layers. *Applied Mechanics Review*, **44**, 1–25.

Ray, G. C. and B. P. Hayden. 1992. Coastal zone ecotones. In *Landscape Boundaries. Consequences for Biotic Diversity and Ecological Flow*, ed. A. J. Hansen and F. di Castri, pp. 403–420. New York: Springer-Verlag.

Redfield, A. C. 1958. The biological control of chemical factors in the environment. *American Scientist*, **46**, 205–221.

Reeves, S. A. and M. B. Usher. 1989. Application of a diffusion model to the spread of an invasive species: the coypu in Great Britain. *Ecological Modelling*, **47**, 217–232.

Refsgaard, J. C. 1996. Terminology, modelling protocol and classification of hydrological model codes. In *Distributed Hydrological Modelling*, ed. M. B. Abbott and J. C. Refsgaard, pp. 17–26. Dordrecht: Kluwer Academic.

Reiners, W. A. 1983. Transport processes in the biogeochemical cycles of carbon, nitrogen, phosphorus and sulfur. In *The Interaction of Biogeochemical Cycles*, ed. B. Bolin and R. Cook, pp. 145–176. New York: Wiley.

1986. Complementary models for ecosystems. *American Naturalist*, **127**, 59–73.

1995. Ecosystem of the Great Plains: scales, kinds and distributions, In *Conservation of Great Plains Ecosystems: Current Science, Future Options*, ed. S. R. Johnson and A. Bouzaher. Dordrecht: Kluwer Academic.

Reiners, W. A. and K. L. Driese. 2001. The propagation of ecological influences across heterogeneous environmental space. *BioScience*, **51**, 939–950.

Reiners, W. A., S. Liu, K. G. Gerow, M. Keller and D. S. Schimel. 2002. Historical and future land use effects on N_2O and NO emissions using an ensemble modeling approach: Costa Rica's Caribbean lowlands as an example. *Global Biogeochemical Cycles*, **16**, 223–240.

Ren, G., D. J. Reddish and B. N. Whittaker. 1987. Mining subsidence and displacement prediction using influence function methods. *Mining Science and Technology*, **5**, 89–104.

Renard, K. G., G. R. Foster, G. A. Weesies and J. P. Porter. 1991. RUSLE: revised universal soil loss equation. *Journal of Soil and Water Conservation*, **46**, 30–33.

Reynolds, J. F. and J. Wu. 1999. Do landscape structural and functional units exist? In *Integrating Hydrology, Ecosystem Dynamics, and Biogeochemistry in Complex Landscapes*, ed. J. D. Tenhunen and P. Kabat, pp. 273–296. Chichester, UK: Wiley.

Rhodes, O. E., Jr and E. P. Odum. 1996. Spatiotemporal approaches in ecology and genetics: the road less traveled. In *Population Dynamics in Ecological Space and Time*, ed. O. E. Rhodes, Jr, R. K. Chesser and M. H. Smith, pp. 1–7. Chicago, IL: University of Chicago Press.

Rhodes, O. E., Jr, R. K. Chesser, and M. H. Smith (ed.) 1996. *Population Dynamics in Ecological Space and Time.* Chicago, IL: University of Chicago Press.

Richards, G. D. 1990. An elliptical growth model of forest fire fronts and its numerical solution. *International Journal of Numerical Methods and Engineering,* **30,** 1163–1179.

1993. The properties of elliptical wildfire growth for time-dependent fuel and meteorological conditions. *Combustion, Science and Technology,* **92,** 145–171.

1995. A general mathematical framework for modeling two-dimensional wildland fire spread. *International Journal of Wildland Fire,* **5,** 63–72.

Richey, J. E., J. M. Melack, A. K. Aufdenkampe, V. M. Ballester and L. L. Hess. 2002. Outgassing from Amazonian rivers and wetlands as a large tropical source of atmospheric CO_2. *Nature,* **416,** 617–620.

Riehl, H. 1978. *Introduction to the Atmosphere.* New York: McGraw-Hill.

Robertson, G. P. 1989. Nitrification and denitrification in humid tropical ecosystems: potential controls on nitrogen retention. In *Mineral Nutients in Tropical Forest and Savanna Ecosystems,* ed. J. Proctor, pp. 55–69. Oxford: Blackwell Scientific.

Rodhe, H. 1992. Modeling biogeochemical cycles. In *Global Biogeochemical Cycles,* ed. S. S. Butcher, R. J. Charlson, G. H. Orians and G. V. Wolfe, pp. 55–72. San Diego, CA: Academic Press.

Rogers, D. 1983. Pattern and process in large-scale animal movement. In *The Ecology of Animal Movment,* ed. I. R. Singland and P. J. Greenwood, pp. 160–180. Oxford: Oxford Science Publications, Clarendon Press.

Rogowski, A. S. and J. L. Goyne. 2002. Modeling dynamic systems and four-dimensional geographic information systems. In *Geographic Information Systems and Environmental Modeling,* ed. B. Parks, M. Crane and K. Clarke, pp. 122–159. Upper Saddle River, NJ: Prentice Hall.

Romme, W. H. 1982. Fire and landscape diversity in Yellowstone National Park. *Ecological Monographs,* **52,** 199–221.

Romme, W. H. and D. G. Despain. 1989. Historical perspective on the Yellowstone fires of 1988. *BioScience,* **39,** 700–706.

Romme, W. H., M. G. Turner, R. H. Gardner *et al.* 1997. A rare episode of sexual reproduction in aspen (*Populus tremuloides* Michx.) following the 1988 Yellowstone fires. *Natural Areas Journal,* **17,** 17–25.

Root, R. B. 1967. The niche exploitation pattern of the blue-gray gnatcatcher. *Ecological Monographs,* **37,** 317–350.

Rothermel, R. C. 1972. *A Mathematical Model for Predicting Fire Spread in Wildland Fuels. Research Paper INT-115.* Ogden, UT: USDA Forest Service Intermountain Forest and Range Experiment Station.

Roughgarden, J., S. D. Gaines and S. Pacala. 1987. Supply side ecology: the role of physical transport processes. In *Organization of Communities: Past and Present,* ed. P. Giller and J. Gee, pp. 491–518. London: Blackwell.

Rundel, P. W. 1981. Structural and chemical components of flammability. In *Proceedings of the Conference: Fire Regimes and Ecosystem Properties (General Technical Report no-26),* ed. H. A. Mooney, T. M. Bonnicksen, N. L. Christensen, J. E. Lotan and W. A. Reiners, pp. 183–207. Ogden, UT: USDA Forest Service.

Russel, E. W. B. 1983. Indian-set fires in the forests of the northeastern United States. *Ecology,* **64,** 78–88.

Ryszkowski, L. 1992. Energy and material flows across boundaries in agricultural landscapes. In *Landscape Boundaries. Consequences for Biotic Diversity and Ecological Flows,* ed. A. J. Hansen, and F. di Castri, pp. 270–284. New York: Springer-Verlag.

Samayoa, A. M., A. P. Thurow and T. L. Thurow. 2000. *A Watershed-level Economic Assessment of the Downstream Effects of Steepland Erosion on Shrimp Production, Honduras. Technical Bulletin 2000–01.* Washington, DC: US Agency for International Development – Soil Management Collaborative Research Support Program.

Sanchez-Pinero, F. and G. A. Polis. 2000. Bottom-up dynamics of allochthonous

input: direct and indirect effects of
seabirds on islands. *Ecology*, **81**, 3117–3132.

Sarmiento, J. L. and C. LeQuere. 1996. Oceanic
carbon dioxide uptake in a model of
century-scale global warming. *Science*,
274, 1346–1350.

Sarmiento, J. L., P. Monfray, E. Maier-Reimer
et al. 2000. Sea–air CO₂ fluxes and carbon
transport: a comparison of three ocean
general circulation models. *Global
Biogeochemical Cycles*, **14**, 1267–1281.

Schimel, D., J. Melillo, H. Tian *et al.* 2000.
Contribution of increasing CO₂ and
climate to carbon storage by ecosystems of
the United States. *Science*, **287**, 2004–2006.

Schippers, P., J. Verboom, J. P. Knaapen and
R. C. van Apeldoorn. 1996. Dispersal
and habitat connectivity in complex
heterogeneous landscapes: an analysis
with a GIS-based random walk model.
Ecography, **19**, 97–106.

Schlesinger, W. H. 1997. *Biogeochemistry. An
Analysis of Global Change*, 2nd edn. San
Diego, CA: Academic Press.

Schnoor, J. L. 1996. *Environmental Modeling. Fate
and Transport of Pollutants in Water, Air, and
Soil.* New York: Wiley.

Schulze, E. D., F. M. Kelliher, C. Korner, J. Lloyd
and R. Leuning. 1994. Relationships
among maximum stomatal conductance,
ecosystem surface conductance, carbon
assimilation rate, and plant nitrogen
nutrition: a global ecology scaling
exercise. *Annual Review of Ecology and
Systematics*, **25**, 629–660.

Schumaker, N. H. 1996. Using landscape
indices to predict habitat connectivity.
Ecology, **77**, 1210–1225.

Seaman, D. E. and R. A. Powell. 1996. An
evaluation of the accuracy of kernal
density estimators for home range
analysis. *Ecology*, **77**, 2075–2085.

Selby, M. J. 1993. *Hillslope Materials and Processes*,
2nd edn. Oxford: Oxford University Press.

Sellers, P. J., Y. Mintz, Y. C. Sud and A. Dalcher.
1986. A simple biosphere (SiB) model for
use within general circulation models.
Journal of Atmospheric Science, **43**, 505–531.

Senft, R. L., M. B. Coughenour, D. W. Bailey
et al. 1987. Large herbivore foraging and
ecological hierarchies. *BioScience*, **37**,
789–799.

Seth, A., F. Giorgi and R. E. Dickinson. 1994.
Simulating fluxes from heterogeneous
land surfaces: explicit subgrid method
employing the Biosphere–Atmosphere
Transfer Scheme (BATS). *Journal of
Geophysical Research*, **99**, 18651–18668.

Sharpe, C. F. S. 1938. *Landslides and Related
Phenomena.* Paterson, NJ: Pageant.

Shepard, P. 1991. *Man in the Landscape*, 2nd edn.
College Station, TX: Texas A & M
University Press.

Shinn, E. A., G. W. Smith, J. M. Prospero *et al.*
2000. African dust and the demise of
Caribbean coral reefs. *Geophysical Research
Letters*, **27**, 3029–3032.

Shreve, R. L. 1968. *The Blackhawk Landslide.* GSA
Special Paper Number 108. Boulder, CO:
Geological Society of America.

Sidle, R. C. 1992. A theoretical model of the
effects of timber harvesting on slope
stability. *Water Resources Research*, **28**,
1897–1910.

Singer, F. J., W. Schreier, J. Oppenheim and
E. O. Garton. 1989. Drought, fires, and
large mammals. *BioScience*, **39**, 716–722.

Siniff, D. B. and C. R. Jessen. 1969. A
simulation model of animal movement
patterns. *Ecological Research*, **6**, 185–217.

Sivinski, J. 1981. The nature and possible
functions of bioluminescence in
Coleoptera larvae. *Coleopteran Bulletin*, **35**,
167–179.

Skellam, J. G. 1951. Random dispersal in
theoretical populations. *Biometrica*, **38**,
196–218.

Sklar, F. H. and R. Costanza. 1991. The
development of dynamic spatial models
for landscape ecology: a review and
prognosis. In *Quantitative Methods in
Landscape Ecology*, ed. M. G. Turner and
R. H. Gardner, pp. 239–288. New York:
Springer-Verlag.

Smith, D. D. and D. M. Whitt. 1957. Estimating
soil losses from field areas of claypan soil.
Soil Society of America Proceedings, **12**,
485–490.

Snow, J. T., A. L. Wyatt, A. K. McCarthy and
E. K. Bishop. 1995. Fallout of debris from
tornadic thunderstorms: a historical
perspective and two examples from
VORTEX. *Bulletin of the American
Meteorological Society*, **76**, 1777–1790.

Solon, J. 1999. Integrating ecological and geographical (biophysical) principles in studies of landscape systems. In *Issues in Landscape Ecology*, ed. J. A. Wiens and M. R. Moss, pp. 22–27. Ft Collins, CO: International Association for Landscape Ecology.

Sommerfeld, R. A., A. R. Mosier and R. C. Musselman. 1993. CO_2, CH_4 and N_2O flux through a Wyoming snow pack and implications for global budgets. [Letter] *Nature*, **361**, 140–142.

Spiekermann, K. and M. Wegener. 2000. Freedom from the tyranny of zones; towards new GIS-based spatial models. In *Spatial Models and GIS. New Potential and New Models*, ed. A. S. Fotheringham and M. Wegener, pp. 45–61. London: Taylor & Francis.

Spitz, K. and J. Moreno. 1996. *A Practical Guide to Groundwater and Solute Transport Modeling*. New York: Wiley.

Sprugel, D. G. 1976. Dynamic structure of wave-generated *Abies balsamea* forests in the northeastern United States. *Journal of Ecology*, **64**, 889–891.

Stallard, R. F. 1992. Tectonic processes, continental freeboard, and the rate-controlling step for continental denudation. In *International Geophysics Series 50*: *Global Biogeochemical Cycles*, ed. S. S. Butcher, R. J. Charleson, G. H. Orians *et al.*, pp. 93–121. San Diego, CA: Academic Press.
   2000. Tectonic processes and erosion. In *Earth System Science: From Biogeochemical Cycles to Global Change*, ed. M. C. Jacobson, R. J. Charleson, H. Rodhe and G. H. Orians, pp. 195–229. San Diego, CA: Academic Press.

Stauffer, D. and A. Aharony. 1992. *Introduction to Percolation Theory*. Washington, DC: Taylor and Francis.

Steinetz, C. (ed.) 1996. *Biodiversity and Landscape Planning: Alternative Futures for the Region of Camp Pendleton, California*. Cambridge, MA: Harvard University School of Landscape Design Report Harvard University Press.

Stephens, D. W. and J. R. Krebs. 1986. *Foraging Theory*. Princeton, NJ: Princeton University Press.

Sterk, G. and A. Stein. 1997. Mapping wind-blown mass transport by modeling variability in space and time. *Soil Science Society of America Journal*, **61**, 232–239.

Stoorvogel, J. J., N. van Breemen and B. H. Janssen. 1997. The nutrient input by Harmattan dust to a forest ecosystem in Côte d'Ivoire, Africa. *Biogeochemistry*, **37**, 145–157.

Strahan, R. (ed.) 1996. *Mammals of Australia*, 2nd edn. Washington, DC: Smithsonian Institution Press.

Streeter, H. W. and E. B. Phelps. 1925. *A Study of the Pollution and Natural Purification of the Ohio River, III. Factors Concerning the Phenomena of Oxidation and Reaeration. Public Health Bulletin No. 146*. Washington, DC: US Public Health Service. Reprinted by US Department of Health, Education and Welfare Public Health Assessment, 1958.

Stull, R. B. 1988. *An Introduction to Boundary Layer Meteorology*. Dordrecht: Kluwer Academic.

Sturm, M., J. P. McFadden, G. E. Liston *et al.* 2001. Snow–shrub interactions in arctic tundra: a hypothesis with climatic implications. *Journal of Climate*, **14**, 336–344.

Sugita, S. 1993. A model of pollen source area for an entire lake surface. *Quaternary Research*, **39**, 239–244.

Sulzman, E. W., K. A. Poiani and T. G. Kittel. 1995. Modeling human-induced climatic change: a summary for environmental managers. *Environmental Management*, **19**, 197–224.

Summerfield, M. A. 1991. *Global Geomorphology*. Harlow, UK: Addison Wesley Longman.

Suntharalingam, P., J. L. Sarmiento and J. R. Toggweiler. 2000. Global significance of nitrous-oxide production and transport from oceanic low-oxygen zones: a modeling study. *Global Biogeochemical Cycles*, **14**, 1353–1370.

Sutton, O. G. 1947. The theoretical distribution of airborne pollution from factory chimneys. *Quarterly Journal of the Royal Meteorological Society*, **73**, 426–436.

Swanson, F. J. and G. W. Lienkaempe. 1978. *Physical Consequences of Large Organic Debris in Pacific Northwest Streams. General Technical Report PNW-69*. Ft Collins, CO: USDA Forest Service.

Swanson, F. J., T. K. Fratz, N. Caine and R. G. Woodmansee. 1987a. Landform effects on ecosystem patterns and processes. *BioScience*, **38**, 92–98.

Swanson, F. J., L. Benda, S. Duncan *et al.* 1987b. Mass failures and other processes of sediment production in Pacific Northwest landscapes. In *Proceedings, Streamside Management: Forestry–Fishery Interactions*, ed. E. O. Salo and T. W. Cundy, pp. 9–38. Seattle, WA: University of Washington.

Swanson, F. J., S. M. Wondzell and G. E. Grant. 1992. Landforms, disturbance, and ecotones. In *Landscape Boundaries. Consequences for Biotic Diversity and Ecological Flows*, ed. A. J. Hansen, and F. di Castri, pp. 304–323. New York: Springer-Verlag.

Tabazadeh, A., E. J. Jensen, O. B. Toon, K. Drdla and M. R. Schoeberl. 2001. Role of the stratospheric polar freezing belt in denitrification. *Science*, **291**, 2591–2594.

Tang, S. M., J. F. Franklin and D. R. Montgomery. 1997. Forest harvest patterns and landscape disturbance processes. *Landscape Ecology*, **12**, 349–363.

Tansley, A. G. 1935. The use and abuse of vegetational concepts and terms. *Ecology*, **16**, 284–307.

Tarrason, L., S. Turner, and I. Floisand. 1995. Estimation of seasonal dimethyl sulphide fluxes over the North Atlantic Ocean and their contribution to European pollution levels. *Journal of Geophysical Research*, **100**, 11623–11639.

Tauber, H. 1965. Differential pollen dispersion and the interpretation of pollen diagrams. *Danmarks Geologiske Undersgelse*, **11**, 69.

Teal, J. M. 1962. Energy flow in the salt marsh ecosystem of Georgia. *Ecology*, **43**, 614–624.

Theis, C. V. 1935. The relation between the lowering of the piezometric surface and the rate and duration of the discharge of a well using groundwater storage. *Transactions of the American Geophysical Union*, **2**, 519–524.

1967. Aquifers and models. In *Proceedings of a Symposium on Ground Water Hydrology*, ed. M. A. Marino, p. 138. Middleburg, VA: American Water Resources Association.

Thom, A. S. 1975. Momentum, mass and heat exchange of plant communities. In *Vegetation and the Atmosphere*, ed. J. L.

Monteith, pp. 57–93. New York: Academic Press.

Thompson, S. A. 1999. *Hydrology for Water Management*. Brookfield, VT: A. A. Balkema.

Thompson, S. L. and D. Pollard. 1995. A global climate model (GENESIS) with a land–surface–transfer scheme (LSX). Part 1: present climate simulations. *Journal of Climate*, **8**, 732–761.

Thornton, P. E. and S. W. Running. 1999. An improved algorithm for estimating incident daily solar radiation from measurements of temperature, humidity, and precipitation. *Agricultural and Forest Meteorology*, **93**, 211–228.

Tianchi, L. 1983. A mathematical model for predicting the extent of major rockfall. *Zeitschrift für Geomorphologie*, **27**, 473–482.

Tilman, D., R. May, R. Lehman and M. Nowak. 1994. Habitat destruction and the extinction debt. *Nature*, **371**, 65–66.

Todd, P. 1996. *Trailing Clouds of Glory. Poems by William Wordsworth*. London: Pavilion Books.

Toffler, A. 1970. *Future Shock*. New York: Random House.

Tovee, M. J. 1995. Ultra-violet photoreceptors in the animal kingdom: their distribution and function. *Trends in Ecology and Evolution*, **10**, 455–460.

Turchin, P. 1998. *Quantitative Analysis of Movement. Measuring and Modeling Population Redistribution in Animals and Plants*. Sunderland, MA: Sinauer.

Turner, B. D. 1994. *Workbook of Atmospheric Dispersion Modeling: An Introduction to Dispersion Modeling*. Boca Raton, FL: CRC Press.

Turner, M. G. and R. H. Gardner. 1991a. Quantitative methods in landscape ecology: an introduction. In *Quantitative Methods in Landscape Ecology: The Analysis and Interpretation of Landscape Heterogeneity*, ed. M. G. Turner and R. H. Gardner, pp. 3–14. New York: Springer-Verlag.

1991b. *Quantitative Methods in Landscape Ecology: The Analysis and Interpretation of Landscape Heterogeneity*. New York: Springer-Verlag.

Turner, M. G., R. H. Gardiner, V. H. Dale and R. V. O'Neill. 1989. Predicting the spread of disturbance across heterogeneous landscapes. *Oikos*, **55**, 121–129.

Turner, M. G., Y. Wu, W. H. Romme and L. L. Wallace. 1993. A landscape simulation model of winter foraging by large ungulates. *Ecological Modelling*, **69**, 163–184.

Turner, M. G., W. W. Hargrove, R. H. Gardner and W. H. Romme. 1994a. Effects of fire on landscape heterogeneity in Yellowstone National Park, Wyoming. *Journal of Vegetation Science*, **5**, 731–742.

Turner, M. G., Y. Wu, L. L. Wallace, W. H. Romme and A. Brenkert. 1994b. Simulating winter interactions among ungulates, vegetation, and fire in northern Yellowstone Park. *Ecological Applications*, **4**, 472–496.

Turner, M. G., R. H. Gardner and R. V. O'Neill. 2001. *Landscape Ecology in Theory and Practice*. New York: Springer-Verlag.

Urquhart, F. A. 1987. *The Monarch Butterfly: International Traveler*. Chicago, IL: Nelson-Hall.

van Bochove, E., N. Bertrand and J. Caron. 1998. *In situ* estimation of the gaseous nitrous oxide diffusion coefficient in a sand loam soil. *Soil Science Society of America Journal*, **62**, 1178–1184.

van Bochove, E., H. G. Jones, N. Bertrand and D. Prevost. 2000. Winter fluxes of greenhouse gases from snow-covered agricultural soil: intra-annual and interannual variations. *Global Biogeochemical Cycles*, **14**, 113–125.

van der Kloet, S. P. 1988. *The genus* Vaccinium *in North America. Publication 1828*. Ottawa: Research Branch of Agriculture Canada.

Van Wagner, C. E. 1977. Conditions for the start and spread of crownfire. *Canadian Journal of Forest Research*, **7**, 23–24.

van Westen, C. J. and M. T. J. Terlien. 1996. An approach towards deterministic landslide hazard analysis in GIS, a case study from Manizales (Colombia). *Earth Surface Processes and Landforms*, **21**, 853–868.

Varnes, D. J. 1958, Landslide types and processes. In *Landslides and Engineering Practice (Highway Research Board Special Report 29, NAS-NRC Publication 544)*, ed.

E. B. Eckel, pp. 20–47. Washington DC: National Academy of Sciences.

1975. Slope movements in the western United States. In *Mass Wasting*, ed. E. Yatsu, A. J. Ward and F. Adams, pp. 1–17. Norwich, UK: Geographical Abstracts.

1978. Slope movement types and processes. In *Landslide Analysis and Control (National Academy of Sciences, Transportation Research Board Special Report 176)*, ed. R. L. Schuster and R. J. Krizek, pp. 11–33. Washington DC: National Academy of Sciences.

Vasconcelos, M. J. and D. P. Guertin. 1992. FIREMAP: simulation of fire growth with a geographic information system. *International Journal of Wildland Fire*, **2**, 87–96.

Vasconcelos, M. J., S. Silva, M. Tome, M. Alvim and J. M. Cardoso Pereira. 2001. Spatial prediction of fire ignition probabilities: comparing logistic regression and neural networks. *Photogrammetric Engineering and Remote Sensing*, **67**, 73–81.

Viessman, W., Jr, G. L. Lewis and J. W. Knapp. 1989. *Introduction to Hydrology*, 3rd edn. New York: Harper and Row.

Vieux, B. E. 2001. *Distributed Hydrologic Modeling Using GIS. (Water and Science Technology Series.)* Dordrecht: Kluwer Academic.

Vieux, B. E. and N. Gaur. 1994. Finite element modeling of storm water runoff using GRASS GIS. *Microcomputers in Civil Engineering*, **9**, 263–270.

Vieux, B. E., F. G. Moreda and Z. Cui. 2001. Arc.water.fea. An ArcView Extension. DEMO Version. Norman, OK: University of Oklahoma, School of Engineering and Applied Science. [Unpublished manual]

Vines, R. G. 1981. Physics and chemistry of rural fires. In *Fire in the Australian Biota*, ed. A. M. Gill, R. H. Groves and I. R. Noble, pp. 129–149. Canberra: Australian Academy of Science.

Vitousek, P. M., H. A. Mooney, J. Lubchenco and J. M. Melillo. 1997. Human domination of earth's ecosystems. *Science*, **277**, 494–499.

Vogl, R. J. 1966. Vegetational continuum. *Science*, **152**, 546.

Volterra, V. 1926. Fluctuations in the abundance of a species considered mathematically. *Nature*, **118**, 558–600.

Walker D. A., J. C. Halfpenny, M. D. Walker and C. A. Wessman. 1993. Long-term studies of snow–vegetation interactions. *BioScience*, **43**, 287–301.

Walsh, S. J., D. R. Butler, T. R. Allen, and G. P. Malanson. 1994. Influence of snow patterns and snow avalanches on the alpine treeline ecotone. *Journal of Vegetation Science*, **5**, 657–672.

Walter, W. G. 1951. An imitation of life. *Scientific American*, **182**, 42–45.

Ward, A. D. and W. J. Elliot. 1995. *Environmental Hydrology*. Boca Raton, FL: CRC Press.

Ward, D. and J. Blaustein. 1992. The role of satisficing in foraging theory. *Oikos*, **63**, 312–317.

Ward, J. V. 1998. Riverine landscapes: biodiversity patterns, disturbance regimes, and aquatic conservation. *Biological Conservation*, **83**, 269–278.

Warrant, E. 2000. The eyes of deep-sea fishes and the changing nature of visual scenes with depth. *Philosophical Transactions of the Royal Society of London*, **355**, 1155–1159.

Waser, N. M. 1986. Flower constancy: definition, cause, and measurement. *American Naturalist*, **127**, 593–603.

Wasser, P. 1977. Individual recognition, intragroup cohesion, and intergroup spacing: evidence from sound playback to forest monkeys. *Behaviour*, **60**, 28–74.

Watson, D. A. and J. M. Laflen. 1986. Soil strength, slope and rainfall intensity effects on interrill erosion. *Transactions of the American Society of Agricultural Engineers*, **29**, 98–102.

Watson, L. 1982. *The Lightning Bird. One Man's Journey Into Africa's Past*. London: E. P. Dutton.

Watt, A. S. 1947. Pattern and process in the plant community. *Journal of Ecology*, **35**, 1–22.

Wear, D. N., M. G. Turner and R. J. Naiman. 1998. Land cover along an urban gradient: implications for water quality. *Ecological Applications*, **8**, 619–630.

Weaver, J. E. and F. E. Clements. 1929. *Plant Ecology*. New York: McGraw-Hill.

Webb, S. W. and C. K. Ho. 1998. Review of enhanced vapor diffusion in porous media. In *Proceedings of the TOUGH Workshop*, pp. 257–262. Berkeley, CA: Lawrence Berkeley National Laboratory.

Wegener, M. 2000. Spatial models and GIS. In *Spatial Models and GIS. New Potential and New Models*, ed. A. S. Fotheringham and M. Wegener, pp. 3–20. London: Taylor & Francis.

West, D. C., H. H. Shugart and D. B. Botkin. 1981. *Forest Succession. Concepts and Applications*. New York: Springer-Verlag.

Wetzel, R. G. 1983. *Limnology*, 2nd edn. Philadelphia, PA: Saunders College Publishing.

Whelan, R. J. 1995. *The Ecology of Fire*. Cambridge, UK: Cambridge University Press.

White, P. S. 1979. Pattern, process and natural disturbance in vegetation. *Botanical Review*, **45**, 229–299.

White, P. S. and S. T. A. Pickett. 1985. Natural disturbance and patch dynamics. In *The Ecology of Natural Disturbance and Patch Dynamics*, ed. S. T. A. Pickett and P. S. White, pp. 3–13. Orlando, FL: Academic Press.

Whitson, T. D., L. C. Burrill, S. A. Dewey *et al.* (ed.) 1996. *Weeds of the West*, 5th edn. Newark, CA: Western Society of Weed Science.

Whittaker R. H. 1953. A consideration of climax theory. The climax as a population and a pattern. *Ecological Monographs*, **23**, 41–78.

Whyte, A. V. T. 1985. Perception. In *Climate Impact Assessment. Studies of the Interaction of Climate and Society*, ed. R. W. Kates, J. H. Assubel and M. Berberian, pp. 403–436. Chichester, UK: Wiley.

Wieczorek, G. F., M. M. Morrissey, G. Iovine and J. Godt. 1999. *Rock-fall potential in the Yosemite Valley, California. Open-file Report 99-578*. Denver, CO: US Geological Survey.

Wiens, J. A. 1992. Ecological flows across landscape boundaries: a conceptual overview. In *Landscape Boundaries. Consequences for Biotic Diversity and Ecological Flow*, ed. A. J. Hansen and F. di Castri, pp. 217–235. New York: Springer-Verlag.

1995. Landscape mosaics and ecological theory. In *Mosaic Landscapes and Ecological Processes*, ed. L. Hansson, L. Fahrig and G. Merriam, pp. 1–26. London: Chapman & Hall.

1999. Toward a unified landscape ecology. In *Issues in Landscape Ecology*, ed. J. A. Wiens and M. R. Moss, pp. 148–151. Ft Collins, CO: International Association for Landscape Ecology.

2000. Ecological heterogeneity: an ontogeny of concepts and approaches. In *The Ecological Consequences of Habitat Heterogeneity*, ed. M. J. Hutchings, E. A. John and A. J. A. Stewart, pp. 9–31. Oxford: Blackwell Science.

2001a. Central concepts and issues of landscape ecology. In *Applying Landscape Ecology in Biological Conservation*, ed. K. Gutzwiller, pp. 3–21. New York: Springer-Verlag.

2001b. Understanding the problem of scale in experimental ecology. In *Scaling Relations in Experimental Ecology*, ed. R. H. Gardner, W. M. Kemp, V. S. Kennedy and J. E. Peterson, pp. 61–88. New York: Columbia University Press.

2001c. The landscape context of dispersal. In *Dispersal*, ed. J. Clobert, E. Danchin, A. A. Dhondt and J. D. Nichols, pp. 96–109. Oxford: Oxford University Press.

Wiens, J. A., B. van Horne and B. R. Noon. 2000. Integrating landscape structure and scale into natural resource management. In *Integrating Landscape Ecology into Natural Resource Management*, ed. J. Liu. and W. W. Taylor. Cambridge, UK: Cambridge University Press.

Wigmosta, M. S., L. W. Vail and D. P. Lettenmaier. 1994. A distributed hydrology–vegetation model for complex terrain. *Water Resources Research*, **30**, 1665–1679.

Wilczynski, W., H. H. Zakon and E. A. Brenowitz. 1984. Acoustic communication in spring peepers: call characteristics and neurophysiological aspects. *Journal of Comparative Physiology A*, **155**, 577–584.

Wiley, R. H. and D. G. Richards. 1978. Physical constraints on acoustic communication in the atmosphere: implications for the evolution of animal vocalizations. *Behavioral Ecology and Sociobiology*, **3**, 69–94.

Williams, J. L. 1975. Sediment-yield prediction with universal equation using runoff factor. In *Present and Prospective Technology for Predicting Sediment Yields and Sources* (*Agriculture Research Service Publication 540*), pp. 244–251. West Lafayette, IN: US Department of Agriculture.

Williams, J., M. Nearing, A. Nicks, E. Skidmore, C. Valentin, K. King and R. Sasvabi. 1996b. Using soil erosion models for global change studies. *Journal of Soil and Water Conservation*, **51**, 381–385.

Williams, M. W., J. S. Baron, N. Caine, R. Sommerfield and R. Sanford, Jr. 1996a. Nitrogen saturation in the Rocky Mountains. *Environmental Science and Technology*, **30**, 640–646.

Williams, M. W., P. D. Brooks and T. Seastedt. 1998. Nitrogen and carbon soil dynamics in response to climate change in a high-elevation ecosystem in the Rocky Mountains, USA. *Arctic and Alpine Research*, **30**, 26–30.

Williams, R. G. and A. J. McLaren. 2000. Estimating the convective supply of nitrate and implied variability in export production over the North Atlantic. *Global Biogeochemical Cycles*, **14**, 1299–1313.

Williamson, M. 1996. *Biological Invasions*. London: Chapman & Hall.

Willson, M. F. 1993. Mammals as seed-dispersal mutualists in North America. *Oikos*, **67**, 159–176.

Wilson, D. E. 1997. *Bats in Question. The Smithsonian Answer Book*. Washington DC: Smithsonian Institution Press.

Wilson, J. P. 1996. GIS-based land surface/subsurface modeling: new potential for new models. In *Proceedings of the Third International Conference/Workshop on Integrating GIS and Environmental Modeling*. Santa Fe, NM. January, 1996. Santa Barbara, CA: National Center for Geographic Information and Analysis. [CD-Rom].

Wischmeier, W. H. and D. D. Smith. 1978. *Predicting Rainfall Erosion: A Guide to Conservation Planning. Agriculture Handbook 537*. Washington DC: US Department of Agriculture.

Wolfenbarger, D. O. 1946. Dispersion of small organisms. Distance dispersion rates of bacteria, spores, seeds, pollen and insects: incidence rates of diseases and injuries. *American Midland Naturalist*, **35**, 1–152.

Woodruff, N. P. and F. H. Siddoway. 1965. A wind erosion equation. *Soil Science Society of America Proceedings*, **29**, 602–608.

Woodruff, N. P. and A. W. Zingg. 1953. Wind tunnel studies of shelterbelt models. *Journal of Forestry*, **51**, 173–178.

Worton, B. J. 1987. A review of models of home range for animal movement. *Ecological Modelling*, **38**, 277–298.

Worton, B. J. 1989. Kernal methods for estimating the utilization distribution in home-range studies. *Ecology*, **70**, 164–168.

Wright, D. and D. Bartlett. 1999. *Marine and Coastal Geographical Information Systems*. Philadelphia: Taylor & Francis.

Yang, C. T. 1995. *Sediment Transport. Theory and Practice*. Princeton, NJ: McGraw-Hill.

York, D. S., T. M. Silver and A. A. Smith. 1998. Innervation of the supranasal sac of the puff adder. *Anatomical Record*, **251**, 221–225.

Young, R. A., C. A. Onstad, D. D. Bosch and W. P. Anderson. 1989. AGNPS: a nonpoint-source pollution model for evaluating agricultural watersheds. *Journal of Soil and Water Conservation*, **44**, 168–173.

Zeil, J. and J. M. Zanker. 1997. A glimpse into crabworld. *Vision Research*, **37**, 3417–1426.

Zingg, A. W. 1940. Degree and length of slope as it affects soil loss in runoff. *Agricultural Engineering*, **21**, 59–64.

Zollner, P. A. and S. L. Lima. 1999. Illumination and the perception of remote habitat patches by white-footed mice. *Animal Behaviour*, **58**, 489–500.

van Zwaluwenberg, R. H. 1942. Notes on the temporary establishment of insect and plant species on Canton Island. *Hawaiian Planter's Record*, **46**, 49–52.

Index